GERALD L. SCHROEDER

SCIENCE

REVEALS

THE ULTIMATE

TRUTH

A TOUCHSTONE BOOK
PUBLISHED BY SIMON & SCHUSTER
NEW YORK   LONDON   TORONTO   SYDNEY

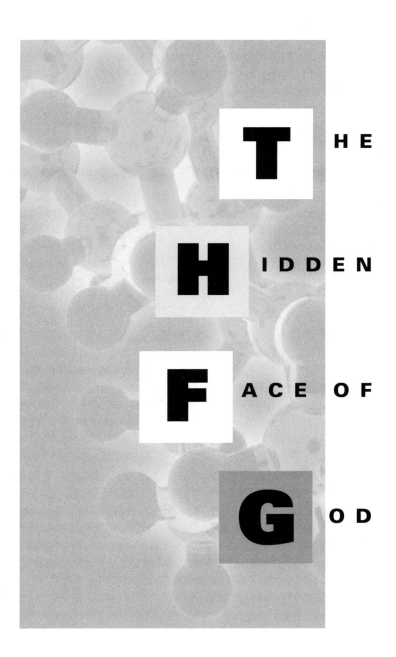

THE HIDDEN FACE OF GOD

TOUCHSTONE
Rockefeller Center
1230 Avenue of the Americas
New York, NY 10020

Copyright © 2001 by Gerald L. Schroeder

First Touchstone Edition 2002

TOUCHSTONE and colophon are trademarks
of Simon & Schuster, Inc.

For information about special discounts for bulk purchases,
please contact Simon & Schuster Special Sales:
1-800-456-6798 or business@simonandschuster.com

Designed by Karolina Harris
Manufactured in the United States of America

20   19   18   17   16   15   14

The Library of Congress has cataloged the Free Press edition as follows:
Schroeder, Gerald L.
The hidden face of God: how science reveals the ultimate truth/Gerald L. Schroeder.
p. cm.
Includes index.
1. Religion and science. 2. God—Proof, Teleological. I. Title.
BL240.2 .S3226 2001
215—dc21   00-050363

ISBN-13: 978-0-684-87059-5
ISBN-10:   0-684-87059-2
ISBN-13: 978-0-7432-0325-8 (Pbk)
ISBN-10:   0-7432-0325-9 (Pbk)

CONTENTS

PROLOGUE: WE ARE THE UNIVERSE COME ALIVE   *xi*

1. THE PUZZLE OF EXISTENCE: AN OVERVIEW   *1*

2. PHYSICS AND METAPHYSICS   *15*

3. THE WORKINGS OF THE UNIVERSE:
   THE PHYSICS OF METAPHYSICS   *25*

4. THE ORDERLY CELLS OF LIFE   *47*

5. MEIOSIS AND THE MAKING OF A HUMAN:
   A STUDY IN SHARED FIDELITY   *70*

6. NERVES: NATURE'S INFORMATION NETWORK   *89*

7. THE BRAIN BEHIND THE MIND   *105*

8. THE PICTURE IN OUR MIND   *129*

9. THINKING ABOUT THINKING: TAPPING INTO THE
   CONSCIOUS MIND OF THE UNIVERSE   *146*

10. ILLUSIONS: GAMES THE BRAIN PLAYS WITH THE MIND   *160*

11. READING BETWEEN THE LINES: WHAT DOES
    IT ALL MEAN?   *173*

EPILOGUE: HINTS OF AN EXOTIC UNIVERSE   *185*

APPENDIX A: DNA/RNA: THE MAKING OF A PROTEIN   *189*

APPENDIX B: MITOSIS AND THE MAKING OF A CELL   *201*

APPENDIX C: MUSCLES: MOLECULES IN MOTION   *211*

ACKNOWLEDGMENTS   *217*

INDEX   *219*

PROLOGUE: WE ARE THE UNIVERSE COME ALIVE

A single consciousness, an all-encompassing wisdom, pervades the universe. The discoveries of science, those that search the quantum nature of subatomic matter, those that explore the molecular complexity of biology, and those that probe the brain/mind interface, have moved us to the brink of a startling realization: all existence is the expression of this wisdom. In the laboratories we experience it as information first physically articulated as energy and then condensed into the form of matter. Every particle, every being, from atom to human, appears to have within it a level of information, of conscious wisdom. The puzzle I confront in this book is this: where does this arise? There is no hint of it in the laws of nature that govern the interactions among the basic particles that compose all matter. The information just appears as a given, with no causal agent evident, as if it were an intrinsic facet of nature.

The concept that there might be an attribute as nonphysical as information or wisdom at the heart of existence in no way denigrates the physical aspects of our lives. Denial of the pleasures and wonder of our bodies would be a sad misreading of the nature of existence. The accomplishments of a science based on materialism have given us physical comforts, invented life-saving medicines, sent people to the moon. The oft-quoted statement, "not by bread alone does a human live" (Deut. 8:3), lets us know that there are *two* crucial aspects to our lives, one of which is bread, physical satisfaction. The other parameter is an underlying universal wisdom. There's no competition here between the spiritual and the material. The two are complementary, as in the root "to complete."

When we see through the camouflage haze that at times convinces us that only the material exists, when we touch that consciousness, we know it. A joyful rush of emotion sweeps over the entire self. This emotional response—some might call it a religious experience—is reported in every culture, from every period. It tells us that we've come home. We've discovered the essence of being. Everyone has felt it at some time or other. Perhaps at a brilliant sunrise, in a work of art, the words of a loved one. The physical and the metaphysical have joined.

If we dared, we'd call the experience spiritual, even Godly. But there's a reluctance to use the "G" word. "Listen to the Force" is acceptable on the great silver screen. If the *Star Wars* scriptwriter had used "Listen to God," the theater would have emptied in a flash. The reluctance is not surprising, considering the bizarre claims erroneously attributed to God through the ages and especially in our age. A bit of scrutiny reveals that most of those claims are based on the expectations for the putative (and generally misunderstood) God of the Bible that we learned as children. Obviously, when our child-learned wisdom is evaluated by the sophistication of our adult minds, that wisdom is bound to seem naive.

The age-old theological view of the universe is that all existence is the manifestation of a transcendent wisdom, with a universal consciousness being its manifestation. If I substitute the word information for wisdom, theology begins to sound like quantum physics. Science itself has rediscovered the confluence between the physical and the spiritual.

If a spiritual unity does underlie physical reality, it would be natural for people to search for that unity. Regrettably in the rush of our daily obligations we often become disconnected, losing the realization that such a unity might actually exist. Our private worlds today seem to expand almost as rapidly as the universe has been physically expanding since its creation. The scientific discoveries facilitating this nomadic mobility of the mind come at a rate that far exceeds the ability of our cultures to adapt. New technologies simply displace old cultural ties, and in doing so jettison traditions that formerly stabilized society.

In the developing world, those referred to as the poorest of the poor are the landless. In a sense we have become landless nomads, cut off from our roots, even in the midst of wealth. We deal in tokens. Other than artists and the one percent of the population that works on a farm, most of us have no relation to the final product of our labors. We buy and sell stocks of companies making products we barely understand. We deal in the ultimate of tokens, money. Money has no intrinsic value. It may promise security, pleasure, even freedom, but it doesn't provide those insatiable and all too often elusive goals. The resulting angst is almost palpable. Divorce rates exceed 50 percent. Violence in homes crosses socioeconomic divisions. Histories have been exchanged for gossamer hopes of a freedom untethered to tradition.

Accessing the consciousness within which we are embedded requires skills that go beyond our intuitions. The amazing, even startlingly illogical, discoveries in physics and biology during the past few decades have given us the tools to gain scientific insight into the metaphysical underpinnings of our world and, in return, acquire spiritual insight into scientific, empirical fact. Understanding nature's wonders need in no way detract from its majesty. By realizing the interwoven complexity of existence, we experience the oneness both by revelation and by reason.

No monk's life of isolated contemplation is being proposed here, no excluding of oneself from the world. The upsurge of interest in meditation, Eastern religions, and kabala reflects an almost desperate search to rediscover our spiritual roots. Those roots are best found while fulfilling the usual responsibilities of adult life, not within some cloister. Exposing the awe of existence *within* the reality of daily life is what this book is about.

We are, each of us, a part of the universe seeking itself. We struggle between a world that seems totally material and the emotional, even spiritual, pull we all feel at times. To relegate, a priori, those feelings of love and joy and spirituality to some assumed function of our ancestors' evolutionary drive for survival masks the greatest pleasure in life, the experiential realization of the metaphysical.

In the following pages, as we journey through the newly discovered marvels of the cosmos, of life, and finally of the brain/mind interface, I ask only that, as you read, you use these facts to reexamine your opinions concerning the origins, evolution, and essence of this wonderful world in which we live.

1

THE PUZZLE OF

EXISTENCE:

AN OVERVIEW

Theology is not simply a matter of interpreting scriptures, be it the Bible, the Koran, or the Tao Te Ching. Theology brings us the amazing concept of a metaphysical Force that brought the physical universe into being. The universe is the physical expression of the metaphysical. All that we know of the putative Creator is found within the physical creation. With this in mind, it is incongruous to describe a theology without the insights of science.

When I picture the earth and solar system hanging in the vastness of space, I feel an anxious need to grab hold of something stable. We're a minuscule speck, somehow floating in the seemingly endless dark of the night sky. It's at those times that I realize all existence hangs by a thread, the breadth of a hair. But the puzzle of our tenuous place in space is secondary to the most baffling riddle of all: that of existence itself. It's a question we might prefer to ignore.

Why is there an "is"? Why is there something, anything, rather than nothing? In our fascination with life's origin and evolution, we bypass this most basic of conundrums. The very fact and nature of existence, the finite aspects of the physical world we view about us, the limited nature of time, space, and matter from which we and all the universe are constructed, force upon us the unsettling reality that at some level there is the metaphysical. Some undefined whatever, transcendent of the physical, produced the physical? Atheist, agnostic, skeptic, and "believer" all share the understanding that some metaphysical non-thing, metaphysical in the sense of being above or outside of the physical, must have preceded our universe or have our universe imbedded in it. That much is a certainty.

But is this metaphysical force that produced our universe Godly? That is not an easy question. It is, however, probably the most important of questions. And the answer really depends on one's definition of God. Just what are the expectations of how an infinite, incorporeal metaphysical creating force would be made manifest within the finite corporeality of the universe? It may be definitions more than reality that separate skeptic from believer, especially if one's expectations of the metaphysical or of the physical are what was learned as a child.

To have a prayer of probing intelligently the subtleties of the metaphysics of existence, we first had better understand the essence of existence at the physical level. In the coming chapters we'll get into the molecular workings that underlie life and discover a complexity so extreme, so overwhelmingly elaborate, that it outdoes science fiction by a league. And then we'll confront a most intriguing enigma: what allows life to conceive a thought, to imagine a rainbow, a symphony? Is the mind totally the neurological workings of the brain or is there a brain/mind interface where the physical mingles with the nonphysical? That would make the brain an instrument, a sort of antenna that taps into the consciousness of the universe. But I'm getting ahead of my story.

Discoveries in the fields of physics, cosmology, and molecular biology during the past few decades have moved us ever closer to what might be termed a metaphysical truth. First of all, there is the creation itself: how did the universe come into existence? And then there is the orderly yet phenomenally complex information we find encoded time and again in all aspects of nature, and most especially in life. The wisdom contained therein is not at all evident in the physical building blocks from which life, even in its simplest of forms, is composed. And finally, there is the puzzle of consciousness and the brain/mind interface. Is mind more that the sum of the physical parts of the brain? The possibility that mind is an emergent property of brain echoes the emergence of wisdom in nature. Of course, the suggestion of a ubiquitous wisdom raises the problem of *imperfect design*. If life is indeed the result of an intelligent force, why are there so many less-than-optimum aspects in its design? I will briefly discuss these topics in this chapter and then pursue the details in the rest of the book. As the saying goes, God is in the details.

Creation and a First Cause

Is there a universal consciousness, an entanglement that at some subtle level ties all existence together? At the moment of the big bang everything, the entire universe, you included, was part of a homogeneous speck—no divisions, no separations—an undifferentiated iota filled with exquisitely powerful energy, a speck no larger than the black pupil of an eye. And before this there was neither time nor space nor matter. That speck was the entire universe. Not a speck within some vacuous space. A vacuum is space. The speck was the entire universe. There was no other space. No outside to the inside of the creation. Creation was everywhere at once. And then the space and the energy stretched out to form all that exists today in the heavens and on the earth. You (and I too) were once part of a total unity, a single burst of radiation. It formed our bodies and led to our thoughts. As incomprehensible as it may seem, we, the essence that was eventually to form our beings, were present at the creation.

When we leave the workings of the macro-world and enter the world of the atom, we are confronted with an even more surprising yet scientifically confirmed picture of our universe.

First we discover that solid matter, the floor upon which we stand and the foundation that bears the weight of a skyscraper, is actually empty space. If we could scale the center of an atom, the nucleus, up to four inches, the surrounding electron cloud would extend to four miles away and essentially all the breach between would be marvelously empty. The solidity of iron is actually 99.9999999999999 percent startlingly vacuous space made to feel solid by ethereal fields of force having no material reality at all. Hollywood would have rejected such a script out of hand and yet it is the proven reality. But don't knock your head against that space. Force fields can feel very solid.

To experience a force field, hold a heavy object out at arm's length. Feel the downward pull. But what is doing the pulling? Gravity, you say. But what produces the gravity? Gravitons of course. But what are gravitons? For that question there is no answer. Where are the strings that tug down toward the earth? They are there, but other than the pull, they are totally imperceptible.

And then quantum physics came onto the scene, with its micro, micro world, only to prove to us that there is no reality. Not even the one part in a million billion that seemed to be solid. Ask a scientist, a physicist, what an electron or the quarks of a proton are made of. She or he will have no answer. Ask the composition of photons—the wave/particle packets of energy that underlie all those micro-particles of matter. The reply will be along the lines of "Huh?"

That all existence may be the expression of information, an idea, a quantum wave function, is not fantasy and it is not some flaky idea. It's mainstream science coming from such universities as Princeton and M.I.T. There is the growing possibility that for all existence, we humans included, there's nothing, nothing as in "no thing," there. In the following chapters we'll discover that the world is more a thought than a thing, more intangible than real. It's essential to see the magnificent subtlety and bounds of how the physical world works if we are going to speculate concerning the metaphysical.

In science, there are easy questions and hard questions. The easy questions, though in practice hard to solve rigorously, all have material solutions. Examples include Newton's phenomenal insight into the laws of motion and gravity; Einstein's revolutionary concepts of relativity. The hard questions relate to what seem to be the intrinsic qualities of nature. Why is there a force of gravity and why does matter generate such a force? Of what are the fundamental particles of matter, electrons, quarks, composed? What is energy?

And then there is what Professor David Chalmers of the department of psychology at the University of Arizona describes as the very hard question. It is that at which I am aiming. What is the mind?

Mind/Matter

The easy part of that question deals with the brain. Wiggle your toes. Feel them? But where do you feel them? Not in your toes. Toes feel nothing. You feel them in your brain. Anyone who has had the misfortune of having a limb amputated can tell you how the missing limb continues to be felt—in the brain. The brain has within it maps of the body that record every sensation and then project that sensation onto the mental image of the relevant body part. But it certainly feels like I'm feeling my toes in my toes. And it is not just the toes. The entire reality, what we see and what we feel, what we smell and what we hear, is mapped in the brain and then those recorded sensations reach out to our consciousness from within the two-to-four-millimeter (about one-eighth-inch) thin wrinkled gray layer, the cerebral cortex, that rests at the top of each of our brains. There is a reality out there in the world, but what we experience—every touch and every sound, every sight, smell and taste—arises in our heads. All our mental images, fantasy or factual, are built on our life's experiences and these are totally physically based. Pure abstraction, even in mathematics, is more than difficult. It is impossible. Even numbers and symbols of calculus take on mental forms.

Through the dedicated and precise work of extraordinarily skilled scientists we now know how and where in the brain each of our sensations is processed and stored. That has been mapped to near perfection. And I explore those maps in the coming chapters. The processes are nothing less that mind-boggling.

And then comes the hard part of the hard question: the sound of music. The waves of sound impinge upon my eardrum and in a beautifully complex path become converted to bio-electrical pulses that are chemically stored in the cortex of my brain. (We'll look at that path in detail.) But how do I hear the sound? Up to and including the storage of the data in the brain, it's all biochemistry. But I don't hear biochemistry. I hear sound. Where's the sound generated in my head? Or the vision; or the smell? Where's the consciousness? Just which of those formerly inert atoms of carbon, hydrogen, nitrogen, oxygen, and on and on, in my head have become so clever that they can produce a thought or reconstitute an image? How those stored biochemical data points are recalled and replayed into sentience remains an enigmatic mystery.

Organization separates living matter from what we perceive as inert, lifeless matter. But there is no innate physical difference between the atoms before and after they organize, learn to take energy from their surrounding environment, and become, as a group, alive. However there is what appears to be a qualitative transition between the awesome biochemistry by which the brain physically records the incoming data and the consciousness by which we become aware of that stored information. In that passage from brain to mind we may be looking for a physical link that does not exist.

Could the consciousness we perceive as the mind be as fundamental as, let's say, the phenomena of gravity generated by mass, or the electrical charge generated by a proton? Gravity and charge are emergent properties. They emanate from substrate particles, as for example gravity from mass, and extend throughout the universe. And though each particle continually emits its field, the emission does not "use up" the particle. For all the force of the field, it seems the force emerges from, but is not made of, the particle from which it arises. Though it originates with the substrate particle, the field is not part of the particle.

Time provides an additional clue in the brain/mind puzzle. We drift in a river of time. There's no possibility of swimming upstream, of going back in time. Destroy every clock, every item that feels the passage of time. The flow of time continues unabated. Time is an intrinsic, ubiquitous quality of our universe, irrespective of whether or not we measure its passage. Might consciousness also be an intrinsic, all-present part of nature, of the universe? In that case every particle would have some aspect of consciousness within. The more complex the entity, the greater would be its awareness of the consciousness housed within.

If quantum physics is correct, and it has an excellent track record of being on the mark, then this suggestion is not so very far from mainstream thought. Every particle, every body, each aspect of existence appears to be an expression of information, information that via our brains or our minds, we interpret as the physical world. Physicist Freeman Dyson, upon his acceptance of the Templeton Prize, stated the idea this way: "Atoms are weird stuff, behaving like active agents rather than inert substances. They make unpredictable choices between alternative possibilities according to the laws of quantum mechanics. It appears that mind, as manifested by the capacity to make choices, is to some extent inherent in every atom. The universe is also weird, with its laws of nature that make it hospitable to the growth of mind. I do not make any clear distinction between mind and God. God is what mind becomes when it has passed beyond the scale of our comprehension." These are the words of one of today's leaders in physics!

From the tiniest grain of sand to the brain of an Einstein, all existence, animate and inanimate, is the product of the same ninety-two elements that are themselves the harvest of the energy of the creation. At every turn an underlying commonality, a unity, emerges from within the diversity.

Now quantum physics has moved us to a more uncanny realization: that at some level this commonality extends to what can be described as an instantaneous mutual awareness among all particles, regardless of the distance of separation. What occurs on one side of the universe seems to affect events on the other side and all points in between. If indeed there is a universal consciousness, this could explain the interrelatedness of particles even when separated by large distances. The most famous demonstration of this is presented in the persistently arcane observations obtained in the "double-slit" experiment. Particles, be they ethereal photons of energy, featherweight electrons or massive atoms, passing through a slit, call it "A," somehow know, and react accordingly, to conditions at far-off slit "B." Among those variable conditions at B is whether or not a conscious observer is present. These data carry a hint that perhaps—just perhaps—consciousness affects the physical manifestation of existence.

It took humanity millennia before an Einstein discovered that, as bizarre as it may seem, the basis of matter is energy, that matter is actually condensed energy. It may take a while longer for us to discover that there is some non-thing even more fundamental than energy that forms the basis of energy, which in turn forms the basis of matter. The renowned former president of the American Physical Society and professor of physics at Princeton University, recipient of the Einstein Award and member of the National Academy of Sciences, John Archibald Wheeler, likened what underlies all existence to an idea, the "bit" (the binary digit) of information that gives rise to the "it," the substance of matter. If we can discover that underlying idea, we will have ascertained not only the basis for the unity that underlies all existence, but most important, the source of that unity. We will have encountered the hidden face of God.

In his book *The Medium Is the Massage* [sic], Marshall McLuhan wrote: "We're not sure who discovered water, but we're pretty sure it wasn't the fish." When one is totally and continually immersed in a single medium, awareness of that unique medium fades.

All we know of ourselves and of the universe, all we know of existence, is what the consciousness of the mind tells us. The one constant aspect of my life is that I am myself. Even in my most fantastic dreams I remain conscious of being me. We swim in a stream of consciousness and accept it without even noticing it.

(Imperfect) Design

When searching for hints of the metaphysical within the realm of the physical, we first must realize that scientists learn to investigate *how* the world works, not *why* it exists. I earned my three academic degrees at the Massachusetts Institute of Technology. My third degree was a Ph.D. in earth sciences and nuclear physics. Ph.D. as in doctor of philosophy. But while with my doctorate of philosophy I learned a great deal in how to probe the functioning of the world, I learned precious little about the philosophy of the world, the study of the why of existence. Until recently science has dealt almost exclusively with a cause-and-effect, materialistic view of reality. The transition from classical physics to quantum physics is only gradually forcing questions of why into the curriculum. The "how" it works is physics. The "why" of existence is already metaphysics.

Another mental hurdle to traverse, and perhaps the most common error in the ongoing, but unnecessary, dispute between science and theology, is the assumption that intelligent design means perfect design. From a human perspective, it's obviously wrong. The *Titanic* in boats, the *Electra* in planes, Three Mile Island and Chernobyl in power generation, all make the point fairly clearly. From a theological stance, the story is the same. Judaism, Christianity, and Islam all draw their understanding of creation from the opening chapters of the Book of Genesis. Genesis claims that a single, eternal, omnipotent and incorporeal God created the universe. This would imply intelligent design on the level of the Divine, and yet from the Bible's view, the universe is far from perfect.

The Bible has broken this sad feature of reality to us by the fourth chapter of Genesis, with Abel's murder by Cain. And remember, according to the text, Abel was the good guy. By chapter four, the good guy is dead. By chapter six, God brings the famous Flood at the time of Noah. From a biblical perspective, very early in the game—in fact, after only six chapters—things had gone from bad to worse and God let them all happen. Society had become so degenerate that God lamented: "I regretted that I made them [all humanity]" (Gen. 6:7). It's at that point that God, in a sense, pressed the reset button and brought on the Flood. The conditions of the world changed and the nine-hundred-year human life-spans that preceded the Flood gradually decreased to the ninety or so years we still experience. Apparently nine-hundred-year longevity was not the optimum design for humans. For the moment, it's irrelevant whether or not you believe that the Flood and all its details actually occurred, or whether or not nine-hundred-year life-spans were possible. What is important here is the biblical message being presented: that the original Divine design of the world was flawed, required Divine retuning, and the Tuner acknowledged this need. According to the Bible, intelligent design, even at the level of the Divine, is not necessrily perfect design.

Biblically, debacle follows upon debacle. Psalm 23 begins "The Lord is my shepherd, I shall not want." This certainly sounds like good news. But just before, we read "My God, my God, why have you forsaken me?" That's Psalm 22.

Opening our eyes to the possibility of imperfect design even at the level of the Divine makes the theological claim of design having been included in the creation much more difficult to dismiss. In fact, biblically speaking, the Designer seems not even to be always visibly active in the accomplishment of the design. As theologically nonkosher as it may sound, at times it seems as if the biblically described Creator is actually feeling Its way along in the unfolding drama of existence. The opening chapter of the Bible, in a brief thirty-one sentences, describes the development of our universe from chaos at its creation to the culminating symphony of life. Seven times in those few sentences we're told "God saw it is good." At the end of the process God was so pleased that the creation was described as "very good." Sounds as if at times it might not have been so good.

What we learn from these biblical events is that intelligent design can be complex design, but it is not necessarily flawless design, even when that design is the work of the Creator. If your image of God is based on a simplistic model of the Divine, don't expect that image to rest easily with the Bible's concept of God or with the real world. Out there on the street, the innocent are often the victims and the guilty at times merely walk away.

There is an ancient source of biblical interpretation, known as kabala, that can help us understand the paradox of what seems to be an occasionally absent God. Based on nuances in the biblical text, the kabala teases out meanings not always apparent in a casual reading. A point of clarification is essential. The popular conception that kabala is mysticism is in error. Kabala is logic, but such deep logic that discovering it can lead to a mystical experience. The kabalistic approach is in essence mathematical. There are two sides to the equation of existence. One side deals with the material world, the other, the spiritual. Any activity to the one brings a parallel activity in the other. Kabala is not the study of God. It would be hubris to think that the finite can comprehend the infinite. Kabala is the study of how the Infinite interacts with the finite creation, what might be called the spiritual physics of the universe.

Two basic concepts govern this mode of study. First, that a

unity pervades and underlies all existence. This is the meaning of "the Eternal is One" (Deut. 6:4).* And then the concept of *tzimtzum*, the Hebrew word for contraction, a spiritual pulling back by the Eternal to yield spiritual space for existence. It is this *tzimtzum* that masks the unity of the creation and allows the slack or leeway in the system. At times the universe seems grandly divine; at times totally natural. Like an Escher painting, with its figure/ground reversals popping back and forth even as we look at it. We struggle to keep one dimension in focus. God, having introduced this level of indeterminacy into the system, allows the world to operate within that range. With humans, it shows itself as our ability to exercise our free will choices. According to the Bible, only when events get way off course does God manifestly step in and redirect the flow. Other than that it all looks natural.

When we piece together these aspects of what the metaphysical might be, we find that the common ground held both by skeptic and believer is actually quite broad. All hold that an eternal non-thing preceded our universe; that our universe had a beginning; that intelligent design, even Divine intelligent design, is not necessarily perfect design. All agree that our universe operates according to laws of great power. One and only one facet of the metaphysical separates the skeptic from the believer. That is whether or not that which preceded creation is immanent and active in the creation. Believer says yes. Skeptic says no. But even on this point, the putative presence of the biblical God can be quite tenuous, and often not evident to the untrained eye.

WE humans like to label things, to wrap our minds around a concept, to define and package it; in essence to limit it so that the concept finds harmony within our human definition of logic.

*Biblical quotations are based on translations from the original Hebrew.

But how does someone label or even think about that which is not part of our physical world? Confining the metaphysical to a physical description totally misses the "meta" aspect.

> And Jacob asked him, and said "Tell me your name." And he said "Wherefore is it that you ask after my name?" (Genesis 32:30)
>
> And [the people] will say to me "What is His name?" What shall I say to them? And God said to Moses, "I will be that which I will be." And He said thus shall you say to the children of Israel "I will be" sent me to you. . . . This is My name *le-olam*. (Exod. 3:13–15)

Le-olam—a Hebrew word with three root meanings: forever, and also hidden, and also in the world. This is my name forever hidden in the world. So how to recognize the presence of the metaphysical?

> On that day [the Eternal] shall be One and Its name One. (Zech. 14:9)
>
> The Eternal is One. (Deut. 6:4)

That is to say, the Eternal *is* One.

But don't think that this is the kind of *one* after which might come the quantities two, three, and four. Nothing as superficial as a number is being revealed in these statements. Rather, the infinite metaphysical as perceived by the physical is an all-encompassing, universal unity. A total oneness is as close as the several trillion neural connections in our brains can come in our quest to discern the infinite.

> You shall know this day and place it in your heart that the Eternal is God in heaven above and on earth below; *ain od*. (Deut. 4:39)

ain od—a Hebrew expression in this verse meaning there is nothing else (compare Deut. 4:39 with Deut. 4:35).

That is to say, there is *nothing else.* Nothing other than this singular totality.

Everything, everything with no exception, is a manifestation of an eternal unity, a transcending ubiquitous consciousness, which many label as God. When you touch that unity, you perceive and also experience the wonder within which you and all the rest of creation are embedded. As the rush of emotion sweeps through your body, your level of consciousness moves from the personal aspect of being self-aware and closes the gulf between the local physical and the universal metaphysical.

This transition in awareness finds parallel in the Bible through our changing relationship with God. At Eden there's a parent-child bond, with the parent (God) setting arbitrary rules—such as don't eat from the tree of knowledge. Later, following the covenant between God and Abraham, a partnership emerges as Abraham and then Moses argue with God over how God should manage the world. And finally, in the Song of Songs, the ultimate relationship is reached, that of a lover (we humans) seeking the loved one (God), and in turn, the loved one seeking the lover. When the encounter finally occurs, we touch and blend with the unity that is all existence. The hidden face of God is to be discovered in that unity.

"The king has brought me into his chambers; we will be glad and rejoice in thee" (Song of Songs 1:4). "All the writings of the Bible are holy," Rabbi Akiva declared, "but the Song of Songs is the holiest of all" (Talmud).

God is not mentioned in the Song of Songs. Even at the ultimate encounter, the face of God remains hidden.

2

PHYSICS

AND METAPHYSICS

If we could know all the properties of all the elements, and every nuance of the laws of nature, we could predict that sodium and chlorine can combine to form sodium chloride, "common" table salt. But could we predict that a collection of atoms can join together to form a brain and then a thought?

System A can produce system B where B is more complex than A. But system A cannot produce a new system, call it B, whose basic parameters are of a type that are totally different from those of A. Try for example to imagine a new type of universe, the form of which has neither time nor space nor matter. Or picture a boundary such as a wall that has an inside edge but no outside edge. Or an existence in which the only dimension is time—no space and no physical items showing the effects of time's passage. All of these systems are outside our range of imagination because all our imaginings are built of the parameters in which we live.

And yet when we observe the sentient wisdom of life emerging from the physical structures of life we are witnessing the emergence of a parameter not evident in those structures.

The classrooms in which I teach introductory and second-level sciences are located in different campus buildings. The five-minute walk from one to the other winds through the alleys of the Old City of Jerusalem, known in Arabic as Al Kuds. It is a spectacular passage. With the turn of a head, you can see two thousand years of history and toil. Here three cultures, Jewish, Christian, and Moslem (chronologically listed), are learning slowly and at times painfully to live together. A while ago, I shared that walk with a colleague, the academic dean of a well-known American university.

After a few comments that skirted the issue, he broached what was uppermost in his mind, though he did it in the third person. "The head of our physics department complains that he wears two hats. During the week it's that of his lab. Comes the weekend, it's the garb of religion. His beliefs have him trapped in a dichotomy. How can one believe in a Creator that is immanently interested in the creation and also believe in the efficacy of scientific investigation, a system based on predicable laws of nature?"

How indeed? In Newton's time it was acceptable to mix physics and metaphysics. Newton himself, though he was among the first to discern the universality of the laws of nature, found no conflict with his firm belief in the God of creation. There was enough mystery left to be explained in the physical world that one could imagine God being active in almost every event.

Scientific progress has changed much of that. During the past three centuries, the mechanisms of many of those "mysterious" phenomena have been fully explained in natural terms. Awe in the finely tuned workings of nature may remain, however much of the mystery is gone. A sunset has lost none of its magnificent beauty, but the ambiguity of what paints the sky red each evening has departed from the experience. We now know it is not the sun or God saying good night to the mountain as the Ulm Uncle told Heidi. It is the preferential scattering of the blue end of the visible light spectrum and the greater penetrating power of the longer red wavelengths that pushes the red photons of light through the atmosphere to the cones of our eyes. (Cones are the receptors in our eye's retina that react chemically and then electrically to the different wavelengths of photons we refer to as light and color.)

Sounds dull, perhaps even confusing to the scientifically uninitiated. Yet knowing the physics steals none of the wonder as the sky turns from blue to crimson, then deep purple, and finally embraces the black of night as Earth's rotation from west to east leaves the sun ever further in the west. While my wife and I are kissing our kids good night in Jerusalem, my nephew is having lunch in Florida. The mystery that remains in the sunset is the riddle of why and how a mixture of seemingly inert, unthinking atoms of carbon, hydrogen, oxygen, and several other varieties can produce humans capable of having the subjective experience we refer to as beauty, or the love that would have us kiss our kids good night. Science is no closer to answering those questions today than it was a century ago.

And if we are watching that sunset from the hills of Jerusalem, we find an even more difficult question: why does the name of this city which has no natural resources, no major industry, no particular strategic importance, not even an indigenous water supply, appear on the front pages of newspapers worldwide more frequently than any city other than Washington, D.C.?

Both theology and science probe the nature of reality. Both seek to find an order in the workings of the world. For theology, it is a given that all reality is imbued with a transcendent spirituality. The surprise of science is that discoveries starting in the early 1900s have moved ever closer to the implication that the world we see about us, the objects in our daily lives that we take for granted as being solid, our bodies included, are expressions or manifestations of something as ethereal as energy. And that below the energy lies information, a totally nonmaterial basis for existence. While not calling this information spiritual, science has significantly closed the gap between the material and the spiritual.

Debating whether or not there is a "ghost in the system," to use the phrase of my colleague Dennis Turner, that gives meaning to sunsets is futile without understanding the system. In the chapters that follow, we'll explore the wonders of the physical world. But beyond the awesome phenomena that we'll discuss lies the most profound puzzle of all. Why is there existence, even of empty space? Even of time? The basic enigma is not whether or not we evolved from an ape, or the ages of rocks. The real question is why is there "being"? The existence of existence is amazing, awesome. We are so much a part of existence that we take it for granted—as a "given" in scientific terms. But step back from the subjectivity and think about it. What caused the big bang? What caused existence? Can we even expect an answer to these questions?

In our arrogance, we humans at times forget we have limits that ultimately constrain what we can know. Far beyond the uncertainty of measurements taught first by the Heisenberg principle of uncertainty, and later amplified to the limits of probability brought by quantum mechanics, there exists a twofold bound to our knowledge.

First the neurological (nerve-based) makeup of our brains may be large, counted even in the trillions of synaptic connections among the individual nerves, with each connection storing a bit of information, but the number is finite and so is our ability to comprehend. A chimpanzee may look at a computer and wonder what it is (if chimps have the ability to wonder). They may realize (if chimps realize) that they are mentally limited; that they do not have sufficient and proper neural connections in their brains to design and build a computer; that there are beings wiser than they and those beings who build the computers are visibly present. To the chimp, there is no question of the existence of the intellectually superior humans that are able to make computers. The chimp knows there is a limit to that which a chimp can comprehend.

We humans don't have that visible cap to our knowledge. There's no life form above us intellectually; nothing to show us that we have cerebral bounds. In theory everything we need to know we can learn. In theory even the sky is not the limit. But that theory, in actuality, cannot be true. Far beyond what the wonders of science may or may not be capable of discovering in the workings of our magnificent universe, there is an unbridgeable limit to our knowledge. That is simply because the amount of information stored in the universe is finite, not infinite. To retain a piece of knowledge, an electron in a computer or in a brain must be elevated to a slightly higher potential. That is, some amount of energy must be added to it. By agreement, that elevation can be indicative of some bit of information. What that particular datum might describe is not important for the present. But what is important is that information storage requires energy, tiny though that energy may be. (Amazingly, our brain's system of information storage uses close to the minimum required energy.) All indications are that our universe is finite. There may be other universes, but ours has its physical limits.

A finite universe has only a finite amount of energy within it. And that means it can store only a finite amount of information, a finite amount of knowledge. To be sure, we can never know all the information therein. It is spread throughout the vast reaches of space. But accessing the totality is not the point so much as the very fact that the totality is limited. It is no wonder then that even as we may write the symbol for infinity within our mathematical equations, we cannot subjectively comprehend—that is, internalize—its meaning. The sky of wisdom has a limit.

Realizing this was for me an epiphany, a qualitative change in my concept of the world's potential. Surround every star in the universe with a host of planets and populate all the planets with billions of Einsteins and Newtons. Let them study and theorize for all eternity and still what they can know will have a predictable limit. A similar limit came with the collapse of determinism. For 150 years the scientific community had a love affair with Laplace's theory of determinism—his erroneous theory that given total knowledge of the entire universe at any given moment one could, using the laws of cause and effect, predict all future events. Then came quantum mechanics, uncertainty, and the fuzzy indefinite boundaries of reality. With those discoveries, determinism evaporated. The cosmos became illogical. Our inability to completely understand our world had been revealed.

Experiments in physics laboratories regularly produce results for which there is no known explanation. *A* proceeds to *C* without passing through *B*. Of course, the operative word here is "known." With undying faith in our powers of reason and the calculating powers of computers, we may in the future discover the mechanism that allows what we might refer to as situation *A* to leap to situation *C* without passing through what now appears to be the essential stage of *B*. There is, however, a growing certainty that dimensions exist in our universe that we cannot sense, regardless of how clever and precise our instrumentation may eventually be. And that the physics of the universe operates within these insensible dimensions as well as within the four spatial measurements of length, width, height, and time that we do experience. Events occurring in those insensible dimensions affect us. In laboratories, we can see the results of experiments but we can't follow the reactions that lead to those results. The paths of those reactions may reside outside the physical measurements of length, width, height, and time. Physics has entered the metaphysical, the realm beyond the physically perceivable, in the fullest sense of that word.

Infinity is not within our reach, neither through the ponderings of philosophy nor the research of the laboratory. But unity is, a unity that encompasses and binds together all existence. The universe is an expression of this oneness.

For many, especially among persons alienated by superficial material aspects of Western society and drawn by default to monochromatic, top-down versions of spirituality, it comes as a surprise that an all-encompassing unity is the core concept not only of science but also of biblical religion. Experiencing that unity is the goal of both disciplines. Oneness is, in fact, the biblical definition of God. Both Judaism and Christianity have discovered the essential theological statement to be "Hear Israel the Eternal our God the Eternal is One" (Deut. 6:4; Jesus as quoted in Mark 12:29). Simply read, the meaning of this biblical verse is that there is one God. But, as we have seen, the oneness stated here is not the integer one after which follow two, three, and four. The Eternal is One teaches a truth far more profound than that there is only one eternal Creator. The One of this verse reveals an Infinity as perceivable by the noninfinite; our experience of an all-encompassing unity.

Monotheism does not limit its claim to there being only one God. Biblical monotheism teaches that everything is an expression of this Unity. "There is nothing else" (Deut. 4:39 as read in juxtaposition to Deut. 4:35). We are intimately a part of the whole. Therefore studying how that whole works must add to the depth of life's experience, much as knowing the politics of the times of Shakespeare adds to our understanding of *Hamlet*. One might conceive of a science without religion, but it is an oxymoron to conceive of religion without science. Revelation and nature are the two aspects of one creation. Theology and science present two versions of that one reality, each version seen from its own unique perspective.

The three religions of Jerusalem claim that humankind is created in the image of God, but they give no description of what that image entails. They even insist that God has no perceptible image. But that means our image of God must not be a thought, since our thoughts are built of images stored within our brains. How does one imagine, or even relate to, an imageless "absolute"?

Perhaps experiencing the Absolute is more a mode of thinking rather than imaging any specific thought. Finding joy in the simple fact of existence could do for starters in that way of thinking. The Bible suggests that we "Worship the Eternal with joy" (Ps. 100). Joy is the biblical sign of reverence. Associating one's self-image and interests with the holistic totality of existence might be the next step. Like it says, "Love your neighbor as yourself" (Lev. 19:18; Matt. 22:37; Mark 12:29). The difficulty is that our world may or may not have a transcendent underlying spirit. But it certainly has the material needs of daily life, and they seem to take priority over our best desires. Spirituality not rooted in the reality of life's material needs rarely effects a meaningful gain.

Professor of kabala Nadine Shenkar describes this bringing of the metaphysical into the material with an exquisite analogy. A tree lifts its leafy branches toward heaven and embraces the blessing of rain. If we didn't know better, we'd expect that the rain enters the tree's leaves and fruit directly from above. It would seem that the growth was from the top down. But not so. The rains must first enter the ground, and only then can they carry nutrients to the leaves and fruits. In this world, reality is rooted in the material. It flows from the bottom up. To reap the beauty of this world, we too must work from the bottom up. The spiritual must first enter the physical before life can flourish. It is no wonder, for example, that the biblical Sabbath is the most physically satisfying day of the week and simultaneously the most spiritual.

There is a range of scientific tools that impact all aspects of life, both physical and spiritual. I'll start with physics, lead into biology, and end with neuroscience, the brain/mind interface, specifically the puzzle of consciousness. Discipline is required at each step, but the reward will be, here and now, the satisfaction of understanding the wisdom that underlies reality. No need to wait for a hoped-for next world or some hypothetical reincarnation for the pleasure of these insights. If there is a Divine plan, reincarnation may or may not be part of that plan. That's up for debate. But for certain, physicality is in the scheme. We have a body. Linking the physical with the spiritual is the goal, and the lesson of the tree.

The pleasureful rush of emotion we experience at the beauty of a sunset touches that link. But something far more grand than the color of the sky ignites our emotions. Buying a few cans of paint at the local hardware store and mixing them to match the red we saw streaming across the evening sky somehow does not give us the same feeling of ecstasy. Viewing a work of art such as Auguste Rodin's *The Burghers of Calais* might also give that pleasure. Even such a totally intellectual experience as discovering a piece of deep wisdom can incite this feeling of awesome wonder. Some common aspect within these diverse experiences transcends the details, resides in the eternal.

In a world made up of such divergent and differentiated entities as stars and pencils and galaxies and people, all of us have felt those rare moments of joy at discovering the undifferentiated whole that lies beneath that complexity. In that moment our sense of individual self dissolves into the immense fabric of the universe. I find a touch of irony in our emotional perception of an all-encompassing unity having a material counterpart in the impersonal laws of physics. Science has revealed that the totality of physical existence is the expression of a single base reality, variegated fields of force, or material-less ringlets of energy each expressing itself in the material variety we see about us. Scientific inquiry of nature has exposed a metaphysical unity. The transcendent beauty we find in a sunset resonates with those physical cosmic roots.

In his famous book, *The Guide of the Perplexed*, the philosopher Moses Maimonides over eight hundred years ago succinctly stated this position: "We must form a conception of the existence of the Creator according to our capacities; that is, we must have a knowledge of metaphysics (the science of God), which can only be acquired after the study of physics (the science of nature); for the science of physics is closely connected with metaphysics and must even precede it in the course of studies. Therefore the Almighty commenced the Bible with the description of the creation, that is with physical science." In Maimonides' time, the idea that science might have something to add to our understanding of spirituality was such an anathema to the religious establishments that his book was burned by Jews and Christians alike!

The light of what we label as Divine is split into two parts. One part is revealed directly. That is the prophetic experience. The other part is hidden in the wisdoms of nature. The era has come when those hidden wisdoms are being discovered, exposing a new and undreamed-of synthesis in what is superficially perceived as a multifaceted and divergent universe.

The two hats worn by my colleague's friend are made of one fabric. They represent one reality but seen from two perspectives. In the following pages, I'd like to investigate both views of our world. I think we will find that Maimonides' claim was well founded. In fact, the Eternal is One.

3

THE WORKINGS

OF THE UNIVERSE

THE PHYSICS OF METAPHYSICS

*As the "clay" of matter is energy, so the building block of energy is
information, wisdom. The universe is the expression of this wisdom.
The universe is the expression of an idea.*

One hundred years ago, a physics professor would have lost
tenure on the spot if caught teaching the concept that matter in
all its forms of solids, liquids, and gases was actually condensed
energy. What hokum it would have seemed. There is conservation of energy and conservation of matter, or so it was believed.
Energy may change form, perhaps from radiant to thermal, or
from kinetic to potential, and so might matter change, from
solid to liquid or gas, but in a closed system the sum total of energy and the sum total of matter remained fixed. That each was
distinct from the other seemed too obvious to be questioned.

Then came Einstein, relativity, and $E = mc^2$. Einstein hypothe-
sized, and it has since been confirmed, that matter, m, intrinsically
represents a specific amount of energy, E. And the type of matter
was immaterial. As bizarre as it seems, a gram of rose petals and a
gram of uranium contain identical amounts of energy. The constant, c^2, in the equation is the speed of light squared or multiplied
by itself. That c^2 is a massive value tells us that even a tiny amount

of matter contains a huge quantity of latent energy. Having personally witnessed the detonation of six nuclear weapons, I suggest we all work and pray for peace. The few grams of mass converted into energy during those tests turned the mountain on which I stood into a quivering Jello-like fluid.

Einstein's discovery of the energy/matter relationship is far greater than merely stating that we can get X kilocalories or Y kilojoules from Z grams of matter. Einstein's insight taught the world that every item, every plant and person and star in a galaxy, is a form of condensed energy, energy in its tangible form. If you had lived all your life in Antarctica and seen only ice and snow, and then were shown a kettle out of which billowed a cloud of steam, could you have believed that the hot vapor was made of the same stuff as the frigid ice, that both were water, but water at different energy levels? We and all we see are frozen energy. If you heat any item far beyond the temperatures that break apart molecules and atoms, ultimately it will revert to pure energy, blending with the radiance of all existence.

Science had discovered the metaphysics within physics, the nonphysical nature of the physical world. In 1900 the German physicist Max Planck had proposed an explanation for the surprising discovery that heated objects radiate light at discrete and fixed energies, rather than as the more logical continuous smear of energies. Aspects of Planck's work were to be essential for the development of Einstein's special laws of relativity, published just five years later. Planck suggested that radiant energy exists only in discrete packages that he called quanta. These quanta of energy, also known as photons, are emitted from a heated object when energized electrons fall from higher to lower orbits.

The implications of this rash idea were immense. If electrons orbiting a nucleus can reside only at specific levels of energy, with no intermediate stages allowed, then how does the electron change from one energy level to another? Not by gradually or even rapidly moving across the divide between orbits. Such a transition would imply that for a finite time, no matter how brief, the electron had an energy intermediate between the higher and the lower orbits. But observations showed that such gradualism was forbidden. The electron simply leaps from one orbit to another in zero time. This step-like transition in energy states is reminiscent of the step-like punctuations in morphology found in the fossil record. Totally illogical as these leaps in nature may be, those are the observed facts. Sometimes nature moves in leaps.

Planck, in short, discovered that the expression of energy at the atomic level is both intricate and nonobvious, even illogical. With Planck's insight, the way was now prepared for quantum mechanics. Building on the concept of the quanta, in 1923, almost a decade after Einstein published his general relativity theory (no longer a theory; now it is a law), the French physicist Louis de Broglie introduced an idea that was even more bizarre in its assertions than Einstein's claim that matter really was a form of energy. De Broglie claimed that "as a result of a great law of nature every bit of energy of proper mass is intrinsically related to a periodic phenomenon of frequency."

In simpler language, all matter had related to it a wavelength and a particular frequency, that is, a certain number of wave cycles per second. Not only had we learned that matter was energy and not "really" matter, we now had to believe that all matter is both particle-like and wave-like. Everything—you and I included—has a wave function. Seventy years of experiments have sustained both Einstein's and de Broglie's preposterous, counterintuitive claims. Absurd though these principles seem to the human mind (which works strongly by deduction—what we have seen to be true in the past should be a good indication of what we will see to be true in the future), this wisdom has made possible transistors and lasers and cellular phones, and even aspects of microbiology. Every piece of electronics that fills our homes, from TVs to microwave ovens, is based on these conceptual, counterintuitive breakthroughs. The universe we have discovered behaves in a manner most illogical. It does not comply with human reason.

The "quantum weirdness" of nature has profound implications. Most significantly it tells us that the world simply is not as it seems. A superficial reading of nature finds differentiation; disparate entities: stars and stones and bottled water and even life and death. At a deeper level that same nature reveals unity. I'm on our balcony. The afternoon Jerusalem sun is filtering through the yellow-green finger leaves on a row of eucalyptus trees planted a century ago to mark the property line. ("So the field of Ephron, . . . and the cave which was therein *and all the trees that were in the field, that were in all the border round about* were made over to Abraham," Genesis 23: 17, 18). De Broglie tells me the leaves and the light are one. Not poetically, though that also, but physically they are one. The fact that he has been proven correct fills me with joy. The universe quietly reveals its unity. God is polite, knocking only gently. We have to listen carefully if we are to hear the report.

THE matter/energy relationship, the quantum wave functions, have profound meaning. Science may be approaching the realization that the entire universe is an expression of information, wisdom, an idea, just as atoms are tangible expressions of something as ethereal as energy. Four basic forces govern the physical interactions of all matter. The first of the four to have been quantified as a force was gravity. On the 28th of April in 1686, Isaac Newton presented his *Principia Mathematica* to the Royal Society of London. The third book contained a description of the universal laws of gravitation, deducing that gravity was intrinsic to all matter. Mass attracts mass throughout the universe in a manner identical to its attraction here on earth, with the force of that attraction being proportional to the masses of matter involved, irrespective of the type. A kilogram of air has the same gravitational effect as a kilogram of steel. Newton's concept of a universal law established a new paradigm, that the universe was really a uni-verse. Uni as in unified.

But what *is* it that produces the force of gravity, that pulls an apple to the ground, or holds the distant moon in orbit? For a force to act at a distance, for the earth to reach out to the moon and hold it, there must be something making contact between the two bodies, otherwise how can the force be felt at a distance? We call that force gravity without really knowing what it is. Anyone who predicts that we are approaching the end of science is fooling himself. The secrets of gravity may one day unlock more human potential than the sum total of the technology we possess today.

For centuries the effects of electrostatic attraction had been observed. Rubbing dry wool did something to the wool (we now know it was scraping off electrons) that drew lint to it, even lint several inches away. Sparks of light darted in the cloth as PJ's were being pulled on, and for some reason the force pulled on our hair. Some strange power was at work.

Magnets presented the same enigma. The magnet is here and the iron nail is there. Empty space lies between. Slowly at first and then with increasing speed, the nail moves toward the magnet. No problem. Everybody knows magnets attract iron. But how? What reaches out from the magnet, beckoning the iron to come hither? It's a lesson in the magic of our universe. Take two horseshoe magnets. Try to push north onto north and south onto south. Look closely at the space between the opposing poles. Nothing is seen, but the force is mightily felt as the poles constantly slip aside, avoiding one another.

By the early 1800s Michael Faraday (1791–1867), the English chemist and physicist, had become intrigued by the seeming impossibility of action at a distance. Something must be carrying the force of the magnet's pull on the nail. Newton, a century and a half earlier, had suffered the same quandary in his work when he postulated the laws of gravity, describing what gravity did, without knowing how it did it or what it was.

The second of the four basic forces was "discovered" two hundred years after Newton. The breakthrough came in 1864, when James Clerk Maxwell was able to integrate the electric forces with the pulling power of a magnet, reducing (or increasing) the two seemingly separate phenomena into one: the electromagnetic force. Maxwell's accomplishment in integrating the two forces was prophetic. It was the first step in a drive to unify all forces—each a separate manifestation of a single unified field.

Both the magnetic and the electrostatic forces are theoretically carried by a single type of entity, photons, totally invisible and massless, observed only in their effects, as iron is drawn to a magnet and lint to wool. But what actually produces the photon in the magnet that reaches out to the iron, or in the nucleus of an atom that sends it hurtling off toward the orbiting cloud of electrons? And what is energy in the first place? We talk of energy so often that we've convinced ourselves it is a real and tangible entity. But that simply is not the case. On close inspection, energy is only a convenient concept for quantifying observed effects. We can't handle a piece of energy. We can store energy in a battery, but that's chemistry. The actual essence of energy remains elusive. Does that move energy closer to the information it may represent than to the matter it can form?

It's the electrostatic force, a force mediated by unseen theoretical photons, that holds you together, just as it does the floor on which you stand. It also keeps you from sliding through the openings in the floor like spaghetti through a colander with oversized holes. The world we see as solid is made solid not by matter, but by ethereal forces carried in photons (themselves a theoretical construct) traveling immense distances between the nuclei and surrounding electron clouds.

I press my hand on a table top. I can't penetrate the wood. But a hammered nail pierces it easily, because the invisible binding forces of the molecules of iron are far stronger than those invisible forces binding the molecules of wood. The blow of the hammer allows the nail to break the bonds created by the invisible photons that produce the very real molecular bonding.

The world of atoms and molecules consists of wavelike particles separated from each other by voids, held in place by never seen, massless photons, traveling at the speed of light among particles that are not only particles but also waves. If you can conceptualize this melee in an intelligible way, I have an urgent suggestion: publish.

If these hypothetical photons traveling between protons in the nucleus and orbiting electrons are what keep the electron cloud on its stable course about the proton-rich nucleus, just how does the photon travel? Travel takes time, yet the photon must work its magic instantly. As the photon leaves the nucleus on its instant journey toward an orbiting electron cloud, how does it know what trajectory to follow? By some estimates, the electron is flying at one tenth of the speed of light. That is quite a clip. Thus the photon must traverse a curved path or somehow "anticipate" where the electron will be so it can make its binding contact. Of course the theoretical physicist might brush these questions aside as being irrelevant, even naive, telling us the electron is a cloud and not a particle. So what holds which part of the cloud in place? The question is relevant and perplexing. To accomplish their tasks, these photons seem to have a great deal of wisdom built into them.

We take these phenomena of gravity and magnetism and electricity for granted. But attempts to understand what produces them has opened a world as bizarre as anything Alice encountered in Wonderland. And this is not fiction.

The four basic forces—gravity, the electrostatic force, and the so-called strong and weak nuclear bonds—have no logical explanation for their existence, but because of them we have a user-friendly universe filled with order and stability, and in at least one location, life conscious of its own existence. As the renowned physicist Freeman Dyson stated so eloquently, "Nature has been kinder to us than we had any right to expect." Consider how the four forces fit together.

1. The strongest of the forces, or the fields these forces produce, is appropriately termed the strong nuclear force. It's what

lets atoms form and keeps them and you from disintegrating into a jumble of protons, neutrons, and electrons—the subatomic particles of which all atoms are composed.

2. The weak nuclear force, some one thousand times weaker than its strong nuclear partner, in a sense works against the binding force within the nucleus and allows for certain types of nuclear disintegration. Sometimes a nucleus expels a particle to increase its overall stability. In doing so it often changes into another element. In some cases an orbital electron is drawn into the nucleus in a process known as electron capture. A reaction takes place between the captured electron (negative charge) and a proton (positive charge) of the nucleus, in which both disappear and a neutron (no charge) appears in their place. Some radioactive transitions are amazing. For example, when an atom of the metal radium expels an alpha particle (two protons and two neutrons), it undergoes a metamorphosis, changing from a metal to a gas known as radon. From solid metal to fluid gas in one step—something to be wary of when you purchase a home. Radon gas can emanate from radium in the structure or from the ground below the house, creating a serious health hazard. Radon, in its decay, pulls the same stunt as did its parent radium. It expels another alpha particle and changes from a gas back into a metal. Both the gas and the metal are composed of the same basic building blocks, protons and neutrons. Such is the wonder that underlies the physics of our universe.

3 and 4. The two remaining forces are the electromagnetic force and gravity. While the former is some one hundred times weaker than the strong nuclear force, gravity is a phenomenal 10^{42} times weaker. That's the strong force divided by a one with forty-two zeros after it. Yet for all its weakness, gravity was the first of the forces to be "discovered" (by Newton) and quantified. That is because of two of its characteristics. Gravity is al-

ways attractive, and it is the only force of the four that acts at large distances. Once beyond the size of an atom, 10^{-10} meters, neither of the nuclear forces is significant. And beyond a few centimeters the same is true for the electromagnetic force. That leaves gravity in charge of most of the space in the universe. It's the force that shaped the structure of the universe, pulling the gases of the early universe into huge galactic clouds, and then squeezing those gases into swirling masses of stars and planets. The sun, the moon, and the earth, when seen from space, are all perfect circles. Gravity did that, pulling the mass of each equally in from all sides, shaping the celestial bodies into spheres.

ONE hundred and three years after Maxwell's brilliant insight that unified the electrostatic and magnetic forces, Abdus Salam and Steven Weinberg proposed a theory unifying the weak and electromagnetic forces. What appear to be disparate particles carrying these fields are actually aspects of the same entity, photons, made manifest at different energies. Once again, beneath the seemingly fractionated nature of existence lies a deeper, powerful idea: that of a unified order.

There is a nuance in the Bible that may portend these discoveries or may just be a coincidence but is interesting nonetheless. We read at the closing of the six days of Genesis, "And God saw all that had been done and behold it was very good" (Gen. 1:31). In the nineteen-hundred-year-old translation of Genesis into Aramaic by the sage Onkelos, the verse is read not as "and behold it was very good," but as "and behold it was a unified order." Unity and order were the stamp of completion. Indeed, unification of all four forces is theoretically possible. Unfortunately, to test any such "final" theory, we would require energies comparable to those present just following the big bang, far beyond the reach of foreseeable technology.

Still, the current Holy Grail of physicists teaches one clear lesson: What superficially appears as diversity is, upon deeper scrutiny, unity. For several years I taught physics for second and third graders in my home laboratory. One of the experiments we did together was to tape strips of colored paper to the head of a centrifuge and then watch them spin. As the speed of the centrifuge increased, we first would see a blur and then, in a snap, the blur turned white as the wavelengths of the colors mixed in our brains. Who would believe that a mixture of every color could possibly result in white? Or conversely, if one had never seen a rainbow or sunlight passing through the prism of a drop of water, could it possibly be that red and green and blue and yellow in all their shades were tucked within the white light of the sun? We live downstream of the prism. We see the world through the prism of creation. Looking back through that prism toward creation yields a most amazing vision, the unification not only of the basic forces of nature, but also of the particles that give rise to those forces. Physics has discovered the oneness of existence.

All our activities must comply with these universal forces of nature. They set limits upon the extent to which we can manipulate matter. The universal forces are themselves confined to act within the parameters of two natural constraints. The Pauli exclusion principle, formulated by Wolfgang Pauli (1900–1958) in 1928, forbids two electrons in an atom or molecule from occupying an exactly equivalent energy state. If this were not the case, electrons in orbit around the nucleus would all soon fall to the lowest level, the orbit closest to the nucleus. Ultradense particles would form with no possibility for the variety of chemical reactions we take for granted.

The second constraint demands that electrons envelop an atomic nucleus at discrete and quantized orbits. These permissible orbits are fixed and constant throughout all nature, each being separated from adjacent orbits by a fixed, given amount of energy. Why this should be an inherent part of our universe is anyone's guess. But if electrons were not confined to specific quanta, if they could assume any and all energy levels, then chemical stability would be a dream unknown, as electrons constantly changed orbits under the influence of the slightest wisp of incoming energy. Molecules would form and disintegrate in a totally unpredictable manner. There would be no chemistry.

Every schoolchild learns the simple model of an atom in which the nucleus is like a sun, with electrons orbiting about it like planets. These two principles of atomic structure combine to make that model possible, with a nucleus composed of protons and neutrons and with electrons in stable orbits about that nucleus. Carbon on earth has identical properties to carbon in the farthest galactic cloud, and all indications are that those properties are extant throughout the universe. This uniformity is what makes chemistry possible, the predictable behavior of atoms reacting with other atoms to form molecules and molecules grouping together repeatedly forming copies of the most amazingly complex products. The DNA of genetic material found in every one of the trillion cells of a newborn baby is but one example of the faithful reproducibility of nature.

The physical world is a wonder-filled phenomenon of unity. The same laws that govern the ten thousand billion billion stars distributed among the hundred billion galaxies of our universe, stretching over some 15 billion light years of space, also govern chemical reactions within the 0.001 centimeter of a cell. From the 10^{-5} meters of an organic cell to the 10^{26} meters of the universe, from the mass of an atom, 10^{-26} kilograms, to the mass of the sun, 10^{30} kg, it's one set of laws. Why? It didn't have to be this way. Why is the universe so intelligible, so consistent? Science alone cannot say. Perhaps we are encountering a hint of the metaphysical held within the physical.

Consider the discovery made by Antoine-Laurent Lavoisier (1743–1794). The French chemist did for chemistry what the English mathematician and physicist, Newton, a century earlier had done for motion. Both appreciated that a single unified set of laws must operate everywhere. Lavoisier realized that all chemical reactions follow paths that are quantitatively reproducible, and which are fixed by the composition of the molecules involved. With this brilliant insight, he laid the basis for all future chemical engineering. After Lavoisier, we could believe that two atoms of hydrogen and one atom of oxygen can combine into H_2O, water, anywhere in the universe.

Unfortunately, the extent of Lavoisier's successes in his younger years was matched by the tragedy of his death. The episode is a lesson in the futility of unbridled spirituality. The French Revolution was in full swing, its leaders filled with the desire to remake the world into a place good for all. It was a sad example of the failure of one-sided spirituality, of a theory run amok. In a bizarre denial of history, E. O. Wilson in *Consilience* tells us that "the Enlightenment thinkers . . . got it mostly right the first time. The assumptions they made [were] of a lawful material world, the intrinsic unity of knowledge, and the potential of indefinite human progress . . . a dream of a world made orderly and fulfilling by free intellect." Yet Wilson writes that the same Enlightenment prepared the ground for the French Revolution's reign of terror in which the state cannibalized its own founders along with France's leading intellectuals. Lavoisier was among them. In 1794, Lavoisier was guillotined. The universe may be intelligible, but we cannot remake it in *our* own image.

Spinoza's ideal in which we might find our way to perfection by rational processes alone is an age-old dream. The nature of the human psyche makes it an unrealistic goal. Both science and religion acting alone have produced irrational and horrific behavior. Like rain in the growth of a tree, the blessing of insight from above must mesh with the roots of material reality from below. Neither by itself can achieve the goal of peace on earth and goodwill toward all.

WE take as givens, as axioms with no apparent explanation, the laws that run the physical universe. They seem completely logical once we know them. But that is because we live with them constantly. Mass attracts mass in direct proportion to the amount of mass present. A kilogram of iron and a kilogram of air have the same gravitational pull. Logical. What else would one expect? It's the mass that engenders the gravitational field.

But must it have been so? Think about the equal but opposite electrical charges of protons and electrons. The former are called, by convention, positive; the latter negative. No problem. That is, no problem until we think a bit deeper. The rest mass of a proton, 1.673×10^{-24} grams, is 1,836 times heavier than the rest mass of an electron. If an electric charge followed the same "logical" law of nature patterned by gravity, then the charge of a proton would be 1,836 times greater than that of an electron. Had that been the case our universe would be a very different place. Huge clouds of electrons, 1,836 times more tightly packed, would circle each atom's nucleus to maintain electrical neutrality. Chemical bonding would be hopelessly weak. Alternatively, electrons might have had the same mass as protons and so the same charge as per the gravity/mass relationship. But heavy electrons would require vastly heavier doses of energy to move them in atomic orbit and to compel the electrons to take part in chemical reactions (chemical reactions are in effect exchanges of electrons among atoms). None of these scenarios makes stable chemistry possible.

What good fortune dictated that charge relationships are not proportional to mass relationships? And if mass, the amount of stuff making up the subatomic particle, does not induce electrical charge, what does? A good question. Perhaps charge is proportional to surface area. Perhaps the particle aspects of protons and electrons have similar surface areas but vastly dissimilar densities. Since we do not have a clue as to their basic composition, anything is possible, even if none of it is logical.

Physicists probe the structure of matter by colliding subatomic particles such as protons and electrons at very high velocities. Analyzing the products of the collisions as the fractured pieces career off in varying directions reveals hints of their constituents. This has opened a Pandora's box. The more deeply matter is probed, the more bizarre it seems. Particles are composed of lesser particles which in turn are composed of still lesser particles, in what has been termed a particle zoo (see Figure 1). The strong possibility exists that at the bottom of the pile we will discover that all the particles are varied manifestations of an underlying energy, which in turn may be the manifestation of something even more ethereal. Call it wisdom, or an idea, information. The Hebrew word integrating all these would be *emet*, reality. It would be the interface between physics and metaphysics. In that case the divide between the physical and metaphysical would be no divide at all. It would be a continuum in which one leads smoothly into the other.

The hint of a flow between the physical and the metaphysical, between the roots of the tree and its branches rich with leaves and fruit, emerges from de Broglie's equating waves and particles,

$$h\nu = mc^2$$

where h is a constant, known as Planck's constant; ν, pronounced *nu*, is the wave frequency (the number of cycles per second) associated with the mass, m, and c^2 is the speed of light squared or multiplied by itself.

Consider the significance of this statement. The left side of the equation describes a wave; the right side a particle. But waves are extended expressions of energy, while particles are discrete entities, having edges and ends, or so we thought in the good old days of logical, classical mechanics. Then came de Broglie and quantum mechanics, and the clear water of reason, the humanly logical universe, became blurred. The wave/particle duality that Einstein had discovered in light, Louis de Broglie extended to matter. There were to be "matter waves." Energy and matter, waves and particles, are all expressions of some deeper reality in which particles and fields of energy and even time blend. If beneath all the weirdness there is logic, a

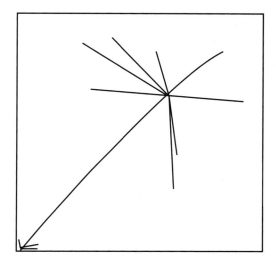

Figure 1
The Particle Zoo
A cloud chamber showing tracks of a single incoming subatomic particle "exploding" into a starlike configuration of nine less massive particles, one of which then "decays" into still lesser particles (lower left corner). Cloud chambers provide visual representation of the actual paths traveled by charged particles as they careen through a chamber in which ultrapure gas saturated with a vapor, such as water vapor, is suddenly cooled to produce supersaturation. As the charged particle passes, it strips electrons from surrounding water atoms, causing the water vapor to condense along its path. In doing so, it produces the track. We have here the opposite of the statement on United States coins: e pluribus unum—one out of many. With subatomic particles, we see e unum pluribus— many out of one! Webster's New Collegiate Dictionary *gives an alternative translation as "one composed of many." This seems to be the reality not only of the U.S.A., but of nature as well.* (Figure after J. Hornbostel and Brookhaven National Laboratory)

thought, a preexisting law, we will have discovered the continuum that links the metaphysical with the physical.

From particle to wave to energy to *idea*. That's the pattern from our perspective, from the inside of creation looking out. The flow is exactly in the opposite direction if we attempt to visualize the path from creation to us.

We experience the physical world of time, space, and matter. That is the information our five senses access. But biblical religion and now physics claim there is more to nature than meets the eye (the ear, nose, etc.). Physics says there are possibly ten, eleven, or twenty-six dimensions. We cannot access all of them, but our inability to do so is not because they are mystical. A bit mysterious they may be, but mystical has nothing to do with the physics that led to these suppositions. One hundred years ago, no J. A. Wheeler would have dared to suggest that all we see about us is actually the expression of condensed information. He'd have been dismissed as a mystic. But then one hundred years ago who would have dreamt that the solid world is really 99.9999999999999 percent empty space made solid by hypothetical, force-carrying, massless particles? And that even that minuscule fraction of matter that *is* matter may not actually *be* matter, but wavelets of energy that we material beings sense as matter?

Metaphysics has entered mainstream, peer-reviewed, university-approved physics, though of course not by that name. In academia it's called quantum mechanics. Modern science has broken the barrier that once separated the illogical from the logical. After 2,350 years, modern physics has lifted itself from the erroneous quagmire of materialist Greek philosophy which promulgated the concept that if you can't touch it or see it, it isn't there. In a sense "heaven" and earth have been joined.

There is even a physics for free will. Quantum mechanics teaches that while the general path of a reaction may be predictable, the exact path is not. There is a probabilistic spread in the path that connects cause with effect. That divergence opens a window of opportunity for what might be choice. It's a topic I'll hold till we get to the functioning of the mind.

PHYSICS has demonstrated that a single substrate underlies all existence. One might ponder: why then is there the diversity seen about us? Differentiation, not simplicity, is the mark of our universe. Solids, liquids, gases, stars, stones, people, trees. According to science, that's simply the way an initial unity played itself out on the stage of the universe. In brief, the scientific scenario reads like this:

The big bang produced, from nothing, a universe composed of photons, energy-packed radiations, unimaginably hot and compressed beyond description, a soup of energy, nearly homogeneous throughout. (What produced the big bang remains a question.) The universe was born as an undifferentiated unity, the oneness science struggles so to rediscover today. But that unity was locked into the ultimate of black holes, a highly compressed mass, so dense with such a huge pull of gravity that nothing, not even something as ethereal as light, could escape. The name black hole arises from the fact that any light passing within the horizon of a black hole is sucked in by the black hole's gravity. With the big bang, the entire universe was such an entity, a near infinite gravitational pull housing and holding everything. That might have been both the beginning and the end of the story. The universe might have remained locked literally within itself. But something happened. We call it inflation, a theory first described by the Russian astrophysicist Alex Starbolinski and a year later by the American Alan Guth.

The universe began to evolve, space stretching, expanding out from a scale (size) that may have been no bigger than a grain of mustard or the black pupil of an eye. What caused this "explosion"? Science cannot say for certain. It was a onetime phenomenon that set the course of history. The map was not exact, but the general direction was set. The laws of nature ensured that, with expansion, galaxies, stars, and planets would emerge. They were, in a sense, preprogrammed into the scheme of things. With each doubling in the scale of the universe, the temperature halved as the waves of radiation were stretched. As the temperature plummeted from an unknown immense value to the relative chill of 10^{13} degrees Kelvin (that's ten trillion degrees!), a small fraction of the energy condensed into the first forms of stable matter, the protons and neutrons that were to eventually compose the entire material world.

Now the nuclear forces took over. Through the alchemy of nature two protons could combine with two neutrons to produce helium, the second lightest of all elements, hydrogen being the lightest element, consisting of a single proton. But that transition to helium requires several steps, as the protons and neutrons join in Lego-like fashion. Fortunately (or as expected?), the joining of a proton and a neutron to form a heavy isotope of hydrogen known as deuterium is sufficiently unstable that it can form only at temperatures below 10^9 degrees Kelvin. That threshold occurred when the universe was a bit over three minutes old.

Alchemy, even universal alchemy, requires heat. By about three minutes the universe had stretched and cooled to a level at which nucleosynthesis could no longer continue. The deuterium instability as well as other instabilities "froze" the universe into a composition having approximately 23 percent helium (by weight) and 77 percent hydrogen, or approximately 7 percent helium (by atoms) and 93 percent hydrogen, with a tiny amount of lithium, the third lightest element, also present. But the major component of the universe, by far, was the photons of radiation. That is very much the universe's composition even today. We live in an expanding mass of energy peppered with a small "contamination" of nuclear particles, about one particle for each ten billion photon rays. We and all we see about us are made of those relatively few, contaminating particles.

After three minutes, the force of gravity became the dominant factor. Building on minuscule inhomogeneities in the expanding universe, differences in density as small as one part in ten thousand, gravity pulled together galactic-sized clouds of hydrogen, shaping them into swirling arrays that often mirrored the exponentially shaped spirals on the shells of snails today, or of soap bubbles as water spins down the drain of a sink.

The material of the universe consisted of hydrogen and helium. But neither hydrogen nor helium is very useful for building something as complex as life. The path taken toward life was elegant. While not great for making life, adequately compressed hydrogen is superb for making stars. And stars are the only sources of long-term stable energy production known in our marvelous universe. Had the deuterium bottleneck of nuclear instability not existed, the very early universe would have made all hydrogen into heavier atoms, possibly iron, with no possibility of ever forming the long-lived stars needed for the patient development of complex life.

Gravity, mass mutually attracting mass, squeezed the swirling gases of hydrogen and helium into clusters of star-sized nebulae, each so tightly packed that temperatures in the cores reached millions of degrees. Hydrogen began to fuse into helium. But the resulting helium weighed a bit less than the components that went into making it. That lost weight represented the mass that had been converted into heat. The first stars were born. (Our sun converts some 660 million tons of hydrogen into approximately 600 million tons of helium every second, with the lost weight given off as the radiation that warms us and lights our days.) In some cases, not all the matter was drawn into the central star's mass. This excess joined to form planets held in orbits around the star.

Within the cores of these newborn stars, the alchemy of nucleosynthesis could proceed at a pace more leisurely than in the hectic first few moments following the big bang. Heavier elements were now being formed. Concurrently, the star's supply of hydrogen fuel was being consumed. Once the supply of hydrogen within a star was exhausted, the outward flow of radiant heat produced by the nuclear reactions in the core ceased. (Every star, in this sense, is a controlled hydrogen bomb.) It was the pressure of this heat streaming outward from the core that had kept the star in equilibrium with the inward pull of gravity. With the heat pressure gone, gravity crushed the star inward from all sides, forcing its collapse, imploding the star upon itself, imploding and then rebounding in the explosion of a supernova. In that act the heavier elements, such as carbon, nitrogen, oxygen, and an array of others, which were needed for complex life, were seeded into the universe.

In time, these newly formed elements could mix with other clouds of the primordial hydrogen, be pulled together, and form into new stars. That is the necessary history of our star, the sun. Our solar system cannot be a first-generation stellar system. There is too much of us in it. Too much of the heavier elements—carbon, nitrogen, oxygen (the sixth, seventh, and eighth of the elements by proton number), and others on up through uranium, number ninety-two. Our sun, the planets, and we ourselves are the products of bygone stars that blasted their contents into space and reformed to make new generations of stars. We are stardust come alive, and somehow conscious of being alive.

That's the scientific scenario. It's all in the laws of physics. Does a metaphysical theology have anything to say about all this? The first few minutes of history just described cannot be found in the Bible, at least not in this detail. So why turn to theology at all?

There are several reasons. First, consider the laws of physics that made it all happen. Did they precede the universe? That would mean laws of physics existed without the physical material upon which to act. Now that sounds a bit bizarre, physics without the physical. But it is the only solution that we scientists can offer.

Some scientists admit having a philosophical problem with the big bang since such a beginning might imply a Beginner. Sir John Maddox, for example, the avidly secular former editor of the journal *Nature*, one of the two most respected peer-reviewed scientific journals, opined that "the big bang is philosophically unacceptable. . . . It is a theory that will be gone in ten years." He wrote that in *Nature*, in August 1989. Maddox's ten years and more have passed. The big bang theory is now more firmly established than ever. Details are uncertain, but the concept that our universe started hot and dense and is expanding out finds ever more supporting data.

The further philosophical problem of there having been a beginning arises with the idea that the beginning of our universe marks the beginning of time, space, and matter. Before our universe came into being, there is every scientific indication that time did not exist. Whatever brought the universe into existence must of course predate the universe, which in turn means that whatever brought the universe into existence must predate time. That which predates time is not bound by time. Not inside of time. In other words, it is eternal. If the laws of physics, or at least some aspect of the laws of physics, did the job of creation, those laws by necessity are eternal.

There are various scientific conjectures currently popular to help explain how a "beginning" happened. One calls for an infinite, eternal medium that can produce universes, one upon the other, somewhat like bubbles rising in an infinite ocean. Each bubble has its own laws of nature. The bubbles with life-supporting laws have life. We are one such bubble. Of course, we can never see outside our bubble to confirm or disprove this theory. Another conjecture is that a field of "virtual energy," a potential, exists. Under specific extreme circumstances that potential can bring virtual particles into reality.

What all of these conjectures have in common is that something, or more accurately stated some non-thing, an eternal whatever, predates our universe. This whatever-it-is has no bodily parts, is totally nonmaterial, is eternal, and though being absolutely nothing physically has the infinite potential to produce vast universes. Sounds familiar. Kind of like the biblical description of God. In fact it *is* the biblical description of God with one significant difference. A "potential field" doesn't give a hoot about the universes it spins off. The Bible, however, claims that the Creator is intimately interested and involved in its creations. Is there any hint of a metaphysical interest in our universe?

Consider the "coincidences" of the first several minutes, and later as planets formed and cooled, coincidences that led to life and consciousness. From a ball of energy that turned into rocks and water, we get the consciousness of a thought. And all by random, unthinking reactions. Even to an atheist, this line of reasoning must seem a bit forced.

This is not a statement of the anthropic principle: "Gee, our universe is so well tuned for life that there must be a Tuner." No, what we see here is far more significant than fine-tuning. We see the consistent emergence of wisdom, of ordered complex information that is nowhere hinted at either in the governing laws of nature or in the particles of matter that form the brain that lies below the mind's thought.

There is a premise commonly applied in physics: Occam's razor, the idea that, all things being equal, the simplest, most elegant explanation tends to be true. A recent book about string theory even used the title *The Elegant Universe*. Why should the universe be elegant? Why should Occam's razor be true? Why are the laws of nature elegant, and from where did they acquire the wisdom to produce intelligent life? Where indeed? Could it be the metaphysical shining through?

THE ORDERLY CELLS

OF LIFE

The opening of the twentieth century marked the onset of the era of physics. Theory and discovery revealed a reality at both subatomic and cosmic dimensions undreamed of just a few decades earlier. Einstein and relativity, Planck and quantum physics, Heisenberg and uncertainty exposed the wisdom within which all existence is embedded.

By the 1950s, especially with the demonstration by Crick, Watson, and Wilkins of the double helix structure of DNA, biological sciences moved to the fore. Almost simultaneously, cybernetics—computer-based information processing—was born. Together with molecular biology, cybernetics brought the world into the era of *information*. For indeed, information lies at the base of both molecular biology and computer science: the former reveals a perplexing depth and breadth of information; the former and the latter both manifest phenomenal ability to manipulate such information.

In the coming chapters we'll take a journey through the wisdom that lies within the cells of life. If scientists had not first discovered such complex processes by which life functions, but merely proposed them as a theory, the scenario of life would be rejected as fantasy. To call the phenomenon of life complex is to trivialize it.

Our study will not be easy. As I discuss the details, some parts will even be tedious. But the wonder of life lies in these details. We'll discern a unifying wisdom embedded in even the simplest forms of life which outshines the wonder of the physics from which it arises. We'll discover that the essence of life, of all life, is the storage, organization, and processing of information. One can only wonder how and from where the complex order of life arose. Life's order is in no way evident either in the atoms and molecules from which that life is composed or in the laws of nature that govern the biochemical interactions among those atoms.

I have no hidden agenda in this tour of the intricacies of life. I make no attempt to deny that life developed from the simple to the complex. Paleontology, biology, and for that matter the Bible each presents its own account of life's flowering. The Bible devotes a mere six sentences to the process. Paleontology records the past in thousands and perhaps millions of fossils. Biology texts fill libraries. Yet all three describe a chain of increasing complexity.

The objective of what follows is twofold: (1) to ponder what processes might have been responsible for life's development in the light of its overwhelming complexity; (2) to discern that the complexity found in life is qualitatively different from that found in the substructures from which it arose. The latter assertion is cause for wonder. Systems can give rise to secondary systems that are more complex than their "parents," but their complexity can be only a fractional extension of that of their parents, an increase in amount but not in type. With life, the increase in complexity from physics to biology seems to be of type as well as amount.

So puzzling is the intricacy of the reactions that power life that at times it seems as if wisdom must be an inherent characteristic of the universe. Our world contains hidden knowledge that is waiting to be expressed. It seems as if a metaphysical substrate is impressed upon the physical.

The Bible, for one, suggests this to be true.

Before dismissing such a suggestion as rubbish, or accepting it as the absolute and obvious truth, let's look at the text. And let's look at it freshly: when a statement is repeated over and over it tends to be accepted whether true or not, and the opening sentence of the Bible has fallen under that spell.

Genesis 1:1 is usually translated as "In the beginning God created the heavens and the earth." Unfortunately, that rendition, which the entire English-speaking world has heard repeatedly, misses the meaning of the Hebrew. The mistake stems from the King James Bible, first published in 1611, based on the Latin Vulgate attributed to St. Jerome in the fourth century and the Greek Septuagint that dates from some 2200 years ago. "In the beginning" is thus three translations downstream from the original.

The opening word, usually translated as "in the beginning," is *Be'reasheet*. *Be'reasheet* can mean "in the beginning of," but not "in the beginning." The difficulty with the preposition "of" is that its object is absent from the sentence; thus the King James translation merely drops it. But the 2100-year-old Jerusalem translation of Genesis into Aramaic takes a different approach, realizing that *Be'reasheet* is a compound word: the prefix *Be'*, "with," and *reasheet*, a "first wisdom." The Aramaic translation is thus "With wisdom God created the heavens and the earth." The idea is paralleled repeatedly in Psalms: "With the word of God the heavens were formed" (Ps. 33:6). "How manifold are Your works, Eternal, You made them all with wisdom" (Ps. 104:24). Wisdom is the fundamental building block of the universe, and it is inherent in all parts. In the processes of life it finds its most complex revelation.

Wisdom, information, an idea, is the link between the metaphysical Creator and the physical creation. It is the hidden face of God.

The human body acts as a finely tuned machine, a magnificent metropolis in which, as its inhabitants, each of the 75 trillion cells, composed of 10^{27} atoms, moves in symbiotic precision. Seldom are two cells simultaneously performing the same act, yet their individual contributions combine smoothly to form life. Gridlock is rarely a problem in the human body.

Ten to the twenty-seventh power—a one followed by twenty-seven zeroes, a thousand million million million million atoms—are organized by a single act when a protozoan-like sperm cell adds its message of genetic material into a receptive egg cell. Combined, these two minuscule cells contain all the information needed to produce the entire body at each stage of its growth, from fetus to adult. We are so embedded in the biosphere that the marvel of its organization has become lost within its commonness.

Until the mid-1970s the accepted wisdom was that the origin of this organization that we refer to as life was the result of chance random reactions among atoms, gradually combining, one chance occurrence building upon another over eons of time until self-replication and then mutation produced the first biological cell. Three billion years were thought to have passed between the formation of liquid water on the formerly molten earth and the appearance of the first forms of life.* That was the message of the fossil record of the time, and that was the logic. After all, how else might one account for the origin of biology? Certainly not by spontaneous generation. Louis Pasteur had laid that primitive idea to rest long ago.

Two to three billion years were available for randomness to do its work. "Given so much time the [seemingly] impossible becomes the possible, the possible probable and the probable virtually certain. One had only to wait. Time itself [and the random reactions able to occur within those eons of time] performs the miracles." So wrote George Wald, professor of biology at Harvard University and Nobel laureate. The article appeared in the August 1954 issue of *Scientific American*, the most widely read science journal worldwide, the Broadway of scientific literature.

---

*All references to ages, such as the age of Earth, age of the universe, and so on are calculated from the time-space coordinates of Earth. From other time-space coordinates, these ages could be vastly different. For a detailed discussion of the age of the universe please see the relevant chapters in my book *The Science of God*.

This speculation over life's origins has within it an important lesson: not to confuse accepted wisdom with revealed fact.

In the mid-1970s came the seminal discovery of Elso Barghoorn. He, like Wald, was at Harvard. Barghoorn assumed correctly that the first forms of life would be small, microbial in size. Using a scanning electron microscope, a tool able to identify minute shapes imperceptible to microscopes that probe images with visible light, Barghoorn searched the surfaces of polished slabs of stone taken from the oldest of rocks able to bear fossils. To the amazement of the scientific community, fossils of fully developed bacteria were found in rocks 3.6 billion years old. Further evidence based on fractionation between the light and heavy isotopes of carbon, a fractionation found in living organisms, indicated the origins of cellular life at close to 3.8 billion years before the present, the same period in which liquid water first formed on Earth.

Overnight, the fantasy of billions of years of random reactions in warm little ponds brimming with fecund chemicals leading to life, evaporated. Elso Barghoorn had discovered a most perplexing fact: life, the most complexly organized system of atoms known in the universe, popped into being in the blink of a geological eye.

If you equate the probability of the birth of a bacteria cell to chance assembly of its atoms, eternity will not suffice to produce one. . . . The speed at which evolution started moving once it discovered the right track, so to speak, and the apparently autocatalytic manner by which it accelerated are truly astonishing. . . . [Yet] chance and chance alone did it all. But it is not, as some would have it, the whole answer, for chance did not operate in a vacuum. It operated in a universe governed by orderly laws and made of matter endowed with special properties. These laws and properties are the constraints that shape evolutionary roulette and restrict the numbers that can turn up. . . . *Faced with the enormous sum of lucky draws behind the success of the evolutionary game, one may legitimately wonder to what extent this success is actually written into the fabric of the universe.* (Emphasis added.)

So wrote Nobel laureate, organic chemist, and a leader in origin of life studies, Christian de Duve, in his excellent book, *Tour of a Living Cell.*

One could say that the ancient and immediate flourishing of life on earth is a miracle—for so it would seem. As a reconciliation between the theological and scientific views of the origins of life, one could also assume de Duve's conclusion that chance events are involved but the system is rigged for life. Chance, luck, and randomness pose no threat to theology when the "chances" and "randomness" are "governed by orderly laws and made of matter endowed with special properties," properties instilled in the universe as the laws of nature at its metaphysical creation. Yet even this totally secular view of our universe provides no simple answer that can explain why the system should be so rigged.

The immediate appearance of life on earth, an event undreamed of prior to its discovery, presses the view of nature as being driven by random reactions into a very tight corner. It is reminiscent of an equally tight corner at the other end of the theological scale, in which persons, by invoking preposterously complex equations, demand that the earth be at the center of the universe despite all data to the contrary (the Bible makes no such claim and even mentions the heavens before the earth).

Interestingly, among the most abundant elements within the material fabric of the universe are those that make up the main components of life. Atoms of hydrogen, carbon, oxygen, and nitrogen account for over 96 percent of the human body. These four, plus helium, are also the most abundant elements in the universe. They are also the only elements that can combine to form the long chains and ring-like molecular structures required for life's processes. The big bang via the laws of nature wove a particularly special fabric.

Not withstanding his first name, de Duve writes in what appears to be a totally secular mode. Even Francis Crick, the avowed "agnostic with a tendency toward atheism," to quote his self-description, described the origin of life as nearly miraculous. De Duve's chemist's view of life—that it is written into the fabric of the universe—resonates with physicist Wheeler's concept that the universe is the expression of an idea. For both Wheeler and de Duve, the evidence stems from discoveries in their respective fields of research, not from verses in the Bible, though these also imply that life is an inherent part of the universal plan. On day three of Genesis, when life is first mentioned, we are told the earth brings forth life. Biblically, the word creation does not appear in relation to the origin of life. No creation means nothing new was required. The Bible claims the necessary components were already present.

But can life be the product only of the laws of nature and the characteristics of the matter upon which those laws operate? Or is there an additional need, the imposition of order from an outside source? Order is known to appear spontaneously in chaotic systems via random reactions. Shake a bag full of letters and occasionally clusters will form that spell words, but never ones that spell long sentences. And further shaking always destroys the initial orderly arrangement. When classical systems are far from equilibrium (shaking letters in a bag is a simple example of such a nonequilibrium system), exotic combinations and reactions among the components multiply. Some of those reactions can, by pure chance, produce orderly arrangements. However, unless this order is somehow locked into place, the system reverts to chaotic disorder. This is the demand of the second law of thermodynamics. In any situation where order is not imposed, momentary order always degrades to chaos. The symphony of life decays to street noise–like chaos once the forces of life cease. Dead bodies decay. Order in life is maintained but only at the expense of large inputs of energy-rich foods and oxygen by which the foods are biologically combusted and the energy within extracted. Life in this sense is a constant drive to maintain order in a universe that favors chaos.

We can find order arising spontaneously in nature, but never on the scale of complexity associated with life. And even here, the order is always the result of a force that imposes it. The beautiful crystals of sodium chloride and bromide along the coast of the Dead Sea (the lowest point on the surface of the earth) are a study in order arising spontaneously. But with crystals, their order is not randomly produced. The electrostatic forces active among the ions of sodium, chloride, and bromide force the atoms to form the regular crystal matrix. The laws of nature predictably produce the crystals.

In 1811, the French physicist Baron Jean Baptiste Fourier derived a mathematical expression describing the propagation of heat in solids by which the flow of heat along a rod is directly proportional to the temperature gradient along that rod. An object's temperature is a measure of the intensity at which its component molecules are vibrating. Touching a piece of hot iron hurts because the iron molecules actually slap your fingers. If the molecular slap is hard enough, it will break the bonds among the molecules of your skin and produce a wound. Fourier's discovery provided a simple law that described how this wave of increased molecular motion progressed through an object. As with crystals, one physical statement describes the orderly behavior of billions of individual molecules.

The earth's rotation from west to east and the permanent low-pressure system in the North Atlantic combine to produce the clockwise flow of the Gulf Stream, which carries warm equatorial water north along the east coast of the United States, then west toward Great Britain and south along Europe's west coast. The interaction of this warm water with the cool air over Britain produces the fog in London. The Gulf Stream represents trillions upon trillions of organized water molecules. They follow a single law, first derived in 1835 by Gaspard de Coriolis, that accounts for the behavior of fluids moving on a rotating earth.

The laws and forces of nature, we discover, are able to impose predictable order on a microscopic as well as on a massive global scale, and are describable in logical mathematical terms. But is the biological order within the living cells of the fish that swim in the Gulf Stream merely a more complex expression of nature's ability to impose physical order, or is there a qualitative difference, a difference in type, between the phenomena? We can predict the formation of crystals of sodium chloride, and even the formation and direction of a Gulf Stream. Could we have forecast the advent of life in an initially lifeless universe? Using a reductionist approach, extrapolating patterns up from basic principles, how far can we go in our divination toward the inception of life?

A truly reductionist argument should start at the beginning. Or even before. But there is no way we could predict, from first principles, a universe with the life-nurturing laws of nature by which we function. So take the existence of our special universe as a given. Accepting the big bang that brought with it these life-friendly laws and the space and energy upon which they act, there'd be no simple logic that would predict that some of this energy would change into stable, lasting matter. We'd know that energy can change into matter. Einstein discovered this aspect of nature and made it famous in his equation $E = mc^2$. But that transition of E into m always produces a pair of particles, matter and antimatter, which then mutually self-annihilate, reverting back to their constituent energy. So in theory the universe should be an expanding ball of ever more dilute (cooler) radiation and no particles of matter. Since we are here, we can surmise that some of the matter, about one part in ten billion, somehow survived the annihilation to form the basic particles, protons, neutrons, electrons, and several others.

So let's move the goalposts yet again, and take our universe, its life-friendly laws of nature, *and* it's basic nuclear particles as givens and start our reductionist approach from there.

Let's say we study the individual subatomic particles, the physical properties of protons, neutrons, electrons. These nuclear particles are the building blocks of atoms, and hence form the basis of all matter. We learn completely their physical characteristics, their rest masses, their relative electrical charges, their natural affinity or repulsion, their wave functions. From this and our knowledge of the laws of nature, which in effect direct interactions among the individual particles, we would predict which combinations would be stable, which not. In doing so we could construct a chart of all possible atoms and the periodic table of elements, in essence the atomic composition of the universe in potential. From the physical conditions just following the big bang, we could determine that of these possible atomic nuclei, the energy of the big bang would produce mostly hydrogen, a single proton, the lightest of elements, and helium, the second lightest of the elements, consisting of two protons and two neutrons. By the end of the period of nucleosynthesis, some three minutes after the creation, we would know that the material of the universe would consist of 93 percent hydrogen atoms, 7 percent helium atoms, plus trace amounts of lithium and beryllium, the third and fourth lightest of the elements. All of this is predictable.

Then, discovering that there were slight variations in the distribution of energy in the early universe, we'd be able to predict that as the universe expanded, gravity would pull these ripples into galaxies and then star-sized clusters. Stars would form, and through the temperature and pressures within the cores of the stars and their subsequent supernovae as the stars exploded, nuclides of the remaining eighty-eight stable elements would form. Gold, atomic number 79, would actually be changed into lead, atomic number 82 (not exactly the dream of an alchemist). A study of the properties of these nuclides would reveal that given the threshold energy of the reaction, sodium and chlorine would combine, sharing an electron to form a stable molecule that might be called salt.

We'd look at two gases, oxygen and hydrogen, and realize that if raised to a threshold temperature, they'd combine in a violent reaction, with two hydrogen atoms sharing electrons with one oxygen atom to produce H_2O, which at room temperature would be liquid. We could decide to call that liquid water. From the properties of these elements we'd predict that the water molecule would have some extraordinary and unique properties. For example, as the water cools to just above freezing and the thermal motion of its V-shaped molecules decreases, they would separate slightly, expanding the water's volume. In doing so, its density would decrease. The result: the solid form, called ice, would be slightly less dense than the liquid from which it froze. Hence ice would float. We could discover that no other common substance would have this property.

This rare characteristic has far-reaching consequences. In large water bodies, such as lakes and oceans, the decrease in density allows the floating ice to serve as a thermal insulator for the warmer water below the ice, in effect limiting the amount of water that can freeze and thus preventing the oceans from becoming solid blocks of ice. The H_2O molecule, we'd see, would have low viscosity in its liquid form, making it an ideal carrier for circulatory systems. An exceptionally large amount of heat would be involved in any change from liquid to solid or vapor, making it an ideal medium for moderating changes in temperature through the heat absorbed upon evaporation as temperature rises and the heat released upon freezing as temperature falls. In short, we could predict that water could be an unusual liquid base useful for a range of complex reactions.

Through this reductionist study of the universe via its atomic structure, we'd even foresee the production of amino acids, combinations of four or five of the lighter elements, with molecules containing up to twenty-three atoms, though from first principles we'd be hard pressed to find any specific benefit in them.

All told, once we take some givens, we can predict much of the chemical world. But that is where our predictions would cease. We can predict all the elements used in life, but there is no indication that we can predict amino acids joining together in chains of hundreds and thousands of units to form proteins and then proteins combining into the symbiotic relationships we refer to as life. When, in 1953, Stanley Miller, then a graduate student at The University of Chicago, produced a few amino acids through purely random reactions among chemicals found naturally throughout the universe, the scientific community felt the problem of life's origin had been solved. Far from it. Subsequent experiments have failed to extend his results. Thermodynamics favors disorder over order. Attempting to get those amino acids to join into any sort of complex molecules has been one long study in failure. The emergence of the specialized complexity of life, even in its most simple forms, remains a bewildering mystery.

That too may be what the Bible tells us. A superficial reading of Genesis chapter one, the creation chapter, marks the end of each of the first six days with "and there was evening and there was morning. . . ." Because the sun is not mentioned until day four, the logic of kabala takes a deeper view of the words. The root meaning of *erev*, the Hebrew word for evening, we learn is mixture, disorder, chaos—just as vision becomes blurry and chaotic at dusk. The root meaning of *boker*, the Hebrew word for morning, is orderly, able to be discerned, as with vision upon sunrise.

The flow of order out of chaos ("And the earth was unformed

and void," Gen. 1:2) is so surprising, so unusual in our universe, that day by day, six times over, the biblical text tells us of this passage to ever more complex arrangements of the existing matter by the seemingly simple statement of "and there was evening and there was morning." Or in the deeper sense, "And there was disorder and there was order." The saga ends with the appearance of human life: "And God saw everything that was made and behold it was very good. And there was evening and there was morning the sixth day" (Gen. 1:31). The millennia-old kabala reads "And God saw everything that was made and behold it was a *unified order.*" Two thousand years ago, the commentators on the Hebrew text saw within the words a wondrous transition from chaos to cosmos, from a jumble of energy and atoms to the dazzlingly complex and unified order of life.

Life beats the odds of chaos over cosmos, but not by defeating the second law of thermodynamics. Nothing does that. Life wins by *outwitting* the second law. The chemistry of a biological cell is the same as the chemistry that forms sodium chloride. One set of rules for all. But unlike sodium chloride, which follows the rules by rote, life has somehow gotten hold of wisdom, of information, that taught it to take energy from its environment, to concentrate that energy, and with it to build and maintain the meaningful complexity of the biological cell.

Cleverly, life scaled the mountain of complexity. What enabled these complex arrangements of carbon plus a few other elements to become so clever remains an enigma. When, as reductionists, we study the individual atoms, we find a sense of choice but no hint of cleverness. Yet somehow, the dust spewed into space by the nuclear furnace of a bygone supernova has become a human brain that learned to make nuclear reactors here on earth. Philosophizing about the origin of our origins, cosmic or biological, may be fascinating, but rarely does it accomplish more than raising other questions. If we are to search the consciousness of the mind in a rational manner, we had best first study the biology of the brain. That biology starts with its most basic unit, the cell.

GOING inside the body and then inside the cell is a journey to wonderland. Enclosed by its outer membrane, a cell's functions are walled off from the outside. When we look at any structure from outside, we get a highly simplified version of its essence. We decide to pick up a pencil and then do so. Not much to it. But in the path leading from the thought to the act, millions of cells and billions of atoms acting on command were required to accomplish that mundane feat. From the outside it seems so straightforward. Like starting a car: just turn the key. Or a computer: just press the power button. A myriad of hours were required to design the circuits and invent the components so that one simple act will activate the billions upon billions of atoms in just the right sequence needed to ignite the motor or light the screen.

If we could see within as easily as we see without, every aspect of existence would be an unfolding encounter with awe; almost a religious experience even for a secular spectator. A biology text presents a diagram, a cutaway view, of a cell. Within the cell, a dozen or so components are shown and labeled—the nucleus, chromatins, cell membrane, ribosomes, and so on. (See Figure 2.) If this were the reality, our metabolism would be one thousandth of what it actually is. A sketch showing all the cellular organelles would be one smear of ink, such is the density of the parts pumping life within.

In the early 1980s, near Hu Bin in the People's Republic of China, I watched the making of a five-hectare fish pond. Thousands of workers lined the banks and swarmed over the ever deepening space. Some broke new earth, some shoveled. A chain of workers passed earth-filled baskets up the bank, where another human chain dumped and returned the emptied bas-

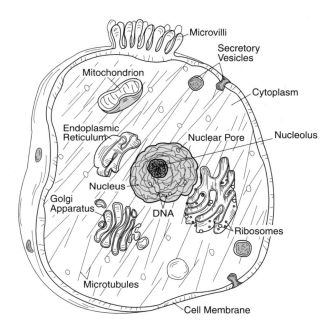

Figure 2

An Idealized Cell with Key Components

Note that only representative organelles are shown. In an actual cell, most of these components are present in the thousands, filling the interior space with activity. The cellular components are not drawn to the same scale. Microvilli, the fingerlike projections, increase the cell's surface area.

kets. The precision was a study in organization. My first encounter with the workings of a living cell was a déjà vu of that experience at the pond. The cell is simply jam-packed, a beehive of activity. Components are stuffed tighter than the circuits of a computer board, and everything is in motion. To get a scale for the rate of activity, consider: on average, each cell in your body, at this second and every second, is forming two thousand proteins. Every second! In every cell. Continuously. And they do it so modestly. For all that activity, we can't feel a bit of it.

A protein is a string of several hundred amino acids, and an amino acid is a molecule having twenty or so atoms. Each cell, every cell in your body, is selecting right now approximately five hundred thousand amino acids, consisting of some ten million atoms, organizing them into preselected strings, joining them together, checking to be certain each string is folded into specific shapes, and then shipping each protein off to a site, some inside the cell, some outside, sites that somehow have signaled a need for these specific proteins. Every second. Every cell. Your body, and mine too, is a living wonder.

The entrance to a living cell is marked by passage through a membrane functioning to keep the bad stuff out, while letting the good stuff in, and expelling what needs to be expelled, waste products and manufactured goods. But who or what decides what comes in and what goes out?

A myriad of portals provide entry, but only if signaled to open and allow entrance. Some of these ports are gated or opened by subtle changes in voltage differences across the membrane. Others open when a molecular key comes and unlocks them, allowing another molecule to pass. The cues come from within the cell if it's a call for the building blocks needed in protein replication, and from outside if, for example, it's a nerve cell coaxing a neighboring cell into action. A vast number of assumptions are woven into the simple act of signaling a membrane port to open. We are so immersed in consciousness, juggling thoughts even when daydreaming, that we project consciousness onto these chemical messengers that control the portals. But where did they get their smarts? Since when do carbon, nitrogen, oxygen, hydrogen, sulfur, phosphorus—the primary building blocks of biology—have ideas of their own, or any ideas at all? They're just atoms strung together to make molecules. Where'd they get the chutzpah to become keepers of the gate?

The laws of nature that govern interactions among atoms are simple and fixed. They produce repetitive formations, such as the crystals at the Dead Sea. Being highly repetitive, they contain little novelty. They do not produce the complex, information-rich molecules we find in life. There is no clue in nature as to how these simple laws could induce the nonrepetitive, multifaceted information we associate with the genetic code or the proteins made from the information stored in those genetic codes.

Basic cell structure is the same throughout the entire biosphere, from the relatively simple structure of a sponge or algal cell to the complexity of a human. A cell's view of life is awash in a sea of water inside and out. The common denominator of all known life is that it is water based. If a cell is to survive, like a wristwatch in a swimming pool, it must be water resistant, but the cell's membrane must be water resistant both inside and out. Nature accomplished this feat by inventing a molecule with a head end that loves water (hydrophilic) and at the other end, two tails that are rejected by water (hydrophobic).

These phospholipids, as they are called, join tail to tail, forming a double-layered membrane, hydrophilic heads facing the water-based cytoplasm inside the cell and the water-based intercellular fluid outside. Sandwiched between inner and outer water-loving surfaces of the membrane are the pairs of hydrophobic tails. The result of this 20 nanometer (billionth of a meter) thick Rube Goldberg contraption is quite remarkable. Though highly flexible, the tenacity of the bonds between the phosopholipid molecules maintains structure. Pinch some skin. It doesn't break or crack. Release it and it returns to the original shape. Puncture a cell membrane with an ultra-sharp needle and then withdraw the needle. Gloop. The membrane reseals the hole and goes on with its work. Because the membrane has both water-loving (polar) and water-rejecting (nonpolar) layers, very few molecules can pass into or out of the cell without explicit permission. Polar molecules face an unfriendly barrier at the nonpolar tails in the middle, while nonpolar molecules can't get past the polar surface.

Such a membrane might make the enclosed cell a dead end. But nature is clever, somehow filled with wisdom. Thousands of receptor and transporter molecules, special proteins, penetrate the wall, determining what can and cannot pass. Muscle cells and especially muscle cells of the heart have large numbers of receptors designed to pass adrenaline, a stimulant hormone that greatly increases a muscle's energy production. At the sensation of danger (sensation did I say? I wonder just which carbon atom is experiencing this emotional trauma?), our reptilian response of fight or flight stimulates the release of large doses of adrenaline into the blood. Taken up by the heart muscles, the beat increases dramatically, pumping oxygen-rich blood to power-hungry muscles in arms and legs. Cells along the small intestine are constructed to absorb glucose, amino acids, and fatty acids, the products of food digestion, and transport these products to the adjacent bloodstream, where they'll be carried to the membranes of cells.

Membrane design is absolutely brilliant. No wonder all biological cells from bacteria at about one micrometer per cell to human cells at 30 micrometers (approximately one thousandth of an inch) function this way. I've been taught that nature did it all.

But there is a catch to this logic of a laissez-faire nature. It is true that lipids and phospholipids can form naturally. And in the presence of water, they do align to form sheets and even spheres. But a leap in information separates a sphere from a cell. That information is the plan of proteins and other molecules required to produce the portals that allow controlled transport across the membrane. Proteins have never been observed to occur naturally, even in a laboratory test tube. Proteins are the products of a cell's metabolism. There's a chicken-and-egg paradox here. Proteins make cells function, but cells are needed to make proteins. The step-by-step formation of an orderly, organized structure such as a protein is far too complex to defeat the second law of thermodynamics—nature's favoring of mayhem over organization—without help. Nurturing environs isolated from the hostile outside world are essential prerequisites. That is just what a cell provides. We're back to the start of the circle.

Yet somehow nature created both these chickens and these eggs in a geological flash after the appearance of the first liquid water on earth.

If we could walk into a cell, our first task would be to keep from getting bowled over. We'd be faced with a myriad of microsized vessels moving in all directions. Something like crossing Broadway against the light. But this is no hodgepodge of motion. Like Broadway there are well-marked traffic lanes. Picture a three-dimensional intersection of several major, multilane highways—crossovers, on and off ramps, an interlacing of cloverleafs one above the other, traffic moving in all directions. Now take it down, with no loss of the complex weave, to a millionth of a meter and repeat it ten thousand times in a sphere 30 millionths of a meter in diameter. You've got an inkling of a biological cell. As a colleague exclaimed when he first viewed a cell: "My God, it's got more connections than the Houston freeway."

The fiber-like roadways, the cytoskeleton, form a sort of jungle gym. They determine the flexible cell shape, hold the internal organelles in place throughout the cell volume, and provide the tracks on which newly manufactured molecules move from point of production to place of need. The item to be moved is "packed" into a pouch-like vesicle covered with thousands of molecular motor proteins. Each motor protein reaches out (extends) a molecular hand, grabs the selected fiber, pulls itself forward by contracting, releases, extends, grabs, and pulls again. Millions of microscopic hands hauling thousands of microsized parcels in every cell, every second of every day.

To meet the energy demanded for every molecular move, every cell has within it the machinery to take glucose from the foods we eat and to combust it, storing the released energy in a power-packed molecule known as ATP—adenosine triphosphate—a sort of biochemical battery that then makes itself available to whatever power-hungry molecule it might encounter. All life uses ATP as its power package, from dandelions to dangerous lions. ATP is nature's global battery.

The subtlety in ATP production is one more lesson in the wisdom held so modestly within life. If you have ever had the misfortune of experiencing a forest fire, flames shooting ten stories into the air, you have seen the power of combustion. To keep this extraordinary potential under control (in order to avoid burning up the cell in capturing energy), nature has found a way to release energy gradually, step by step, in a series of complex biochemical reactions, rather than leaping directly to the end products of carbon dioxide and water as happens in a forest fire. The cleverness of this elaborate bio-series is simply awe inspiring.

Let's follow the path of an ingested bit of carbohydrate. It's like being at a track and field meet, there are so many activities going on simultaneously. Digestion in the mouth and stomach degrades the large carbohydrate molecules into more manageable–sized glucose, which is able to pass through the intestine walls, diffuse into the adjacent bloodstream, and be swept by the blood flow to some glucose-hungry cell. But glucose is a highly polar molecule and so it cannot cross the nonpolar barrier of the cell membrane. Along comes the glucose carrier protein we call insulin, which attaches itself to the glucose. With insulin on site, the membrane gates fly open and the glucose enters the cell. Here, in a complex multistep process in which a dozen intermediate molecules are formed, free-floating enzyme proteins change the glucose into a substance called pyruvate and then, within the cell's many sausage-shaped mitochondria organelles, oxidize the pyruvate into the end products, carbon dioxide and water plus copious amounts of energy-rich ATP. The first two stages of this process are energy intensive and so require an energy source to power them. This is supplied by none other than—you guessed it—ATP (see Figure 3).

Wait a minute! Something's out of order here. This whole process is designed to make energy-rich ATP, and yet just to get the process started we need ATP. It's the chicken-and-egg all over again.

The "waste" products, carbon dioxide and water, must be dealt with. Water is already the medium of life, both inside and outside the cell, so that waste is not a toxic problem. And carbon dioxide is, as we all know, a gas, and therefore far easier to excrete than if it had been a solid as are most metallic oxides at body temperature. This fact, that CO_2 turns out to be gaseous rather than the usual solid oxide, is just another piece of good fortune that we have the choice of taking for granted or, instead, feeling it as a joyful part of the wonderful weave of life; as part of the extraordinary unified order that manifests itself throughout our uni-verse.

One basic cell structure, one basic energy source, one set of organelles common to all life. And one system for regulating this unity, the DNA-RNA team that takes individual lifeless raw

materials and organizes them into living, thinking, choosing beings. The complexity in the commonness stretches the imagination. In his book *Life Itself*, Francis Crick was *almost* on target
when he observed that the origin of life seems *almost* miraculous.

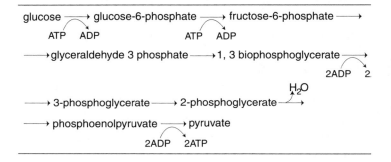

Figure 3

Glucose Metabolism for the Production of Energy-Rich ATP

*The stages of glucose to pyruvate require no oxygen, use two energy-rich mole-
cules of ATP, and produce four ATP molecules, a net gain of two ATP molecules.
This releases approximately 10 percent of the energy within the glucose. The
next stage, the oxygen-rich Krebs cycle, omitted here because of its complexity
(as if the pyruvate process shown here were not itself a mental wrestling match),
releases the remaining energy by reducing the pyruvate to CO_2 and H_2O. The
Krebs cycle takes place within the mitochondria of the cell. One of the many*

molecules essential to the chemical reactions in glucose metabolism, cy-
tochrome C, looks in diagram like this:

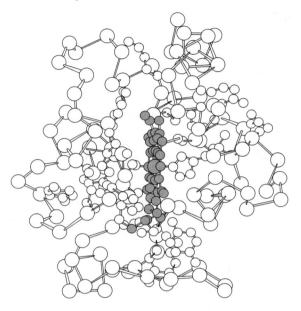

The shape of the molecule is essential for its functioning. With such structural
complexity, one wonders how the cell can tell one molecule from another. And
this is just one example of the dozens of enzymes, all equally complex, all essen-
tial, and all performing just the single act of getting energy from food. That en-
ergy must then be used to facilitate the functions of life. The implications of this
complexity are worth pondering.

MEIOSIS AND THE

MAKING OF A HUMAN

A STUDY IN SHARED FIDELITY

This is the story of metamorphosis. In fact it is the story of your and every human's metamorphosis. If it were a screenplay, we'd say it was fantasy, incredible. But it is reality—it is your history. It is the story of cells dividing and producing daughter cells that take paths very different from that of their parents.

If a caterpillar could speak, could it tell you that it was about to become a butterfly? And if you had not seen it happen already, would you believe it? Who would believe that from eight identical cells clustered together in the shape of a mulberry could come a human?

We've studied the making of the universe and the structure of a biological cell. Let's see what it takes to make a human out of those cells.

After life itself, sex may be the greatest invention ever. It has done wonders for engendering the variety found in the biosphere. And yet for all the variety, the most common type of cellular activity in all life, mitosis, is a process that works against variety. With mitosis, a single cell, upon signal, duplicates all of its organelles, including its package of genetic material, the DNA. It then separates the paired products and divides to yield two identical daughter cells, each a replica of the parent cell.

With sexual reproduction, rather than a single cell acting as the originator of the progeny, a pair of cells unites to produce a daughter cell that is similar but not identical to either of the parent cells. To accomplish this, the new cell has a mix of the genes taken from each parent. And in that mix lies the potential for vastly diverse combinations within each species. Instead of a world full of identical clones, sex produces the wonderful variety that is so much of the spice of life.

But mixing genes from two cells poses a technical problem. Each parent cell has a full complement of genetic material, the DNA. For sex to work, and for the progeny to share contributions from both sides of the bond, each of the parents must be willing to relinquish half of her or his DNA, in technical terms changing from diploid to haploid. The process by which the DNA load is halved is termed meiosis.

Meiosis uses the biotechnology developed for mitosis, but at a few key stages it applies that technology differently. In the following description of the process, it may seem as if a human mind is functioning here. Don't lose sight of the fact that it is all molecules at work. There is no brain on site. The wisdom is built-in.

At the chromosomal level, the events of meiosis are similar for male and female, though the timing is vastly different. Each of the cells, the egg and the sperm, that will be active in reproduction, having duplicated its DNA and therefore holding a double complement of chromosomal material, undergoes mitotic division. The duplicated DNA chromosomes align along a central plane of the cell (see Figure 4). In standard mitotic fashion, each parent cell produces spindle fibers that, with their wonder of synchronous motion, reach out from opposing walls of the cell, attach to each of the pairs of duplicated DNA chromosomes, and pull the pairs apart, separating the chromosome sister pairs into two sets of twenty-three pairs. Just what taught the cells to invent and train these fantastically clever fibers remains to be discovered. The sets are now sequestered on opposite sides of the cell. A motor protein then forms, encircles the outer membrane of the cell with a molecular loop, and proceeds to draw in the loop, pinching the cell into two daughter cells. It's hard to keep in mind that it's all just molecules—no brain power directs these events. The two daughter cells, we discover, are not genetically identical. That is because genes devoted to the same function, for example, eye color or ear size, may carry out that function differently in each of the parents. My wife's eyes are green. Mine are brown. Though within each of our five children both eyes are the same color, among our five kids there's a range of eye colors. Why? Because in the separation and division of the chromosomes, genes for eye color are distributed randomly between the two new cells.

This random placing makes possible a multiple of variations not only in eye color. In excess of five million combinations are possible in the complete set of traits housed within the human genome. In addition to this multiplicity, while the pairs are aligned at the equator of the cell, prior to their being pulled to opposite sides, they swap gene pieces in a process termed crossing over. Each of the two new chromosome sets now contains pieces from the mother and from the father, even though there is only one full set of DNA data in each set. By this stage, the number of possible combinations runs into the multiples of trillions. With this vast potential for variety, it is not surprising that no two humans are identical. With all this genetic promiscuity, it's amazing that the vast majority of births produce normal looking kids.

One of the puzzles of meiosis is the altruistic nature of the cell. Why should a cell willingly give up half of its chromosomal information, and thereby essentially guarantee that its progeny will not be an identical copy of itself? I would have thought that altruism stops at self-destruction. A parent's mixing its chromosomes with those of another is, in a sense, self-destruction, since

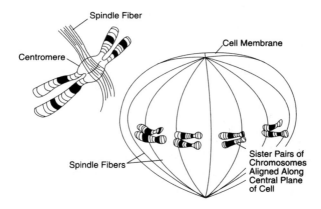

Figure 4

Metaphase in the Cycle of Meiosis

Spindle fibers, composed of microtubules, have attached to the sister pairs of chromosomes at their point of attachment (the kinetochore, or centromere). The spindle fibers will now pull the pairs apart and move each chromosome to an opposite pole of the cell. The process in nothing less than an exquisite ballet played out at the molecular level.

the parent will not be reproduced in the child. Not a very selfish way for a potentially selfish gene to act. The answer lies in the greater wisdom of nature. There is a payoff for the species. Variation aids in species survival by inducing a variety of traits within each species. With variety comes the potential for a wider range of responses to unpredictable environmental changes. But mutations and therefore evolution take place at the level of the individual, not at the level of the species. The individual gains nothing by meiosis. The origin of sex thus remains an unsolved puzzle.

Now comes a further nuance of meiosis. The next cell division in an egg occurs only years later, upon its being fertilized, while in the sperm it occurs immediately. In this division, chromosome replication does not occur. The four new cells (two from each of the previous pair) have only half the complement, twenty-three singular (not paired) chromosomes. These consist of twenty-two plus an X or a Y. Meiosis is complete. Let the cell's sex life begin!

The act of conception is as complex and intriguing as is the meiosis that allows it to occur.

Sperm cells, containing the meiosis-produced single set of chromosomes, are continually produced in the male from puberty through old age. Not so the female eggs. The basis for all the eggs a woman will ever have are already present in the embryo. Following their first meiotic division they wait in limbo until puberty and then fertilization. At birth, some two million pre-eggs are waiting in the ovarian wings, each with a double set of chromosomes—forty-six homologous pairs. These, surprisingly, are not all viable and by puberty only one in five, or some four hundred thousand, remain. At each ovulation, many eggs rupture, but usually only one egg matures. Because of this extravagantly profligate usage of the egg supply, by the time a woman reaches her fifties, the number of remaining eggs is insufficient to maintain the monthly cycle of ovulation.

At ovulation, a follicle in the wall of one of the two ovaries bursts open and a mature egg is released. The previously dormant egg, housing a double set of chromosomes, divides, yielding two cells each with twenty-three pairs (the normal amount). The nuance here is that rather than dividing the cytoplasm equally between the two cells, almost all the cytoplasm and organelles are concentrated in one of the daughter cells, that which is to become the mature egg. The "starved" cell disintegrates. If the egg is fertilized, this excess supply of intracellular material will serve good purpose, providing nutrients and cellular organelles during the early cellular divisions of the embryo.

Along with the released egg, the follicle burst also discharges a cloud of hormones, protective cells that surround and accompany the egg, and a chemical attractant, a sort of perfumed agent that appears to act as a lure to coax sperm to egg. These all are swept into the adjacent fallopian tube, the 10-centimeter channel that winds a path from ovary to uterus. Cilia lining the fallopian tube speedily brush the perfumed come-on ahead of the egg, through and out the fallopian tube, into the uterus. The massive egg lags behind, moving at a much more leisurely pace.

Now for the activity at the other end. As soft fingerlike projections and muscular contractions of the fallopian tube urge the egg along toward the uterus, sperm ejaculated by the male into the woman's vagina start their passage toward the egg. The egg's journey from ovary to uterus takes about four days. The life of an unfertilized mature egg is less than a day, about ten to fifteen hours. Clearly we cannot wait for the egg to reach the uterus if fertilization is to be successful. The sperm fortunately are up to the task. The life of a sperm is about two days, and they are built for traveling (see Figure 5).

Recall that, unlike the mature egg, sperm at the time they are released into the vagina contain only a half set of genetic material. This is concentrated in its head, along with chemicals needed later for penetration of the egg. While the egg is a thousand times bigger than most human cells, comparable in size with the period at the end of this sentence, the head of the sperm is among the smallest of cells, 2 to 3 millionths of a meter (microns) in diameter, 5 microns long. The rest of the sperm is a pure powerhouse designed for motion, a motorized tail, mechanistically a near replica of the cilia that coat our lungs and the flagella of a protozoan, though at some 50 microns in length it is far longer than a protozoan flagellum. Its midsection is wrapped with mitochondria, producers of the energy-rich ATP molecule. These will supply the copious amounts of biological fuel needed to power the flagellum muscle's furious whiplike lashing that propels the sperm during its long journey. Interest-

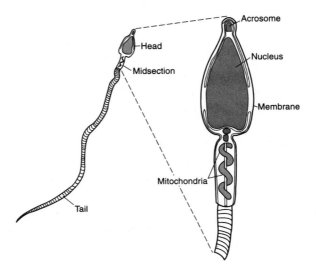

Figure 5

Diagram of a Sperm Cell (Spermatozoon)

The term "spermatozoon" is well chosen since the sperm, with its long flagellum—or tail—is very much akin to a protozoan. The head, some 2 to 3 micrometers in diameter and about 5 micrometers long, houses the DNA-containing nucleus and the acrosome, which contains the enzymes needed to penetrate the egg's wall upon contact. The entire head is covered by a plasma membrane. Mitochondria in the midsection supply the ATP needed to power the whiplike actions of the flagellum as it propels the sperm forward at 3 to 5 millimeters per minute. The tail is greater than ten times the length of the head. With these dimensions, the sperm is truly built for traveling.

ingly, while almost all cells use glucose for the mitochondria energy source, sperm uses fructose.

Somehow sensing the perfumed lure that was released with the egg, the sperm swim from the vagina, across the uterus, into and up the correct fallopian tube. Moving at three millimeters per minute, sperm can reach the egg within a few hours, well under the fifteen-hour life span of the egg.

Approximately three hundred million sperm enter the vagina. They die in droves as they traverse the chemically hostile environment of the fallopian tube. But though hostile, these same harsh chemicals strip the sperm head of its protective coating, preparing it to penetrate the egg. Of the few hundred sperm that finally reach the egg, only one can enter. All batter the egg's protective coating, releasing protein enzymes that work at stripping the egg of its protective coating till one can touch the actual egg wall. A chemical key, a specially shaped protein on the sperm's head, must match the chemical lock, another specially shaped protein on the egg. No foreigners are allowed; a wrong species' sperm would waste an egg. Only if the match is proper can the sperm force its way in, and fertilization occur. A new potential life is under way.

Entrance of the sperm stimulates calcium ion channels in the egg's wall to open. This immediately closes the egg's membrane to all other invaders. One is company. Two's a crowd, genetically speaking. The closure is first induced by a change in the electrical charge across the membrane, and then by a release of a chemical hardener that cross-links the gelatinous outer membrane coating.

The sperm is now within the egg, but it is kept at bay. The egg still has a full set of chromosomes, twenty-three pairs. Half must be discarded. At this stage the egg undergoes meiotic division, division without chromosome replication. Though the egg's chromosomes are divided equally between the two new cells, as before, almost all the cytoplasm is concentrated in one daughter cell, that which contains the sperm's nucleus. This done, the sperm and egg each duplicate their respective twenty-three chromatids, motor proteins move the two nuclei together, the nuclear membranes open, and the chromatids mingle, each finding and pairing with its corresponding partner. Spindle fibers and motor proteins then pull one member of each pair to opposite poles, two nuclear membranes form, and mitotic cellular division occurs. Another bio-ballet, a coordinated dance of molecules within the body, has passed. A few days and several further mitotic divisions will go by before this new bit of life reaches the uterus.

At this point, one has to bend over backward to accept that all of these necessary and interwoven steps have evolved randomly. There are only two forms of reproduction, mitosis and meiosis. There are no intermediate forms visible in nature. Yet somehow, according to evolutionists, a batch of lucky mutations allowed meiosis to blossom on the tree of life.

In human reproduction, meiosis gives way to mitosis, but with an amazing twist to the mitotic principle of daughter cells being replicas of the parent. As the cells divide, some of the daughters discover or interpret how and where to become the cells of a heart, and some a nose or toe. How this miracle of structuring occurs remains a speculative mystery. The knowledgeable differentiation among the newly forming cells appears in principle to be orchestrated by concentration gradients of specific molecules within the cluster of cells, but the wonder of it remains. Who or what supplied the scheme? Wisdom is encoded in the very stuff on which it must act, the blueprint and the builder all in one.

Keeping in mind that the identical DNA is now held in each of the cells, the paradox of differentiation is compounded when we realize that the process appears to be accomplished with no prompting from the mother's body. Fetal development identical to that observed in utero is obtained in vitro through four days post-fertilization with human egg and sperm, and with other vertebrates for much larger fractions of total gestation. Somehow the egg knows what it must become.

By the second day after fertilization, three mitotic divisions have yielded a total of eight cells, though with no significant increase in the overall size of the cluster. Until now the cell divisions have been accomplished through enzymes and organelles already present in the egg prior to fertilization. No new materials have been manufactured. The sequestering of most of the cytoplasm and organelles in only one of the two cells of mitosis now plays a key role. This hoard of material provides the base for all initial activity. Until now the genes of the male have played no part.

Prior to the next mitotic division, eight into sixteen cells (on day three), the organelles within each cell are pulled toward the exterior surface of the eight-celled ball. Then in near perfect simultaneity the eight cells divide. The outer daughter cells receive most of the cytoplasm and associated intracellular structure. But unlike the earlier asymmetric divisions where the "starved" daughter cells were destined for death, the inner hungry cells are to be the building blocks of the fetus, and then months later, the baby.

It's day four. The journey down the fallopian tube is about to be completed. The minuscule embryo, now consisting of between thirty-two and sixty-four cells, enters the uterus, releases an enzyme that dissolves the protective shell within which it has traveled since fertilization, and produces and releases a hormone, human chorionic gonadotropin (hCG), that tells the mother's body she is mothering an embryo. With this message comes a command to terminate her menstrual cycle—an absolutely essential step lest in menstruation the nurturing environs of the uterus be expelled. The appearance of hCG in a woman's urine or blood is the first indicator of pregnancy.

This entire cellular juggling act is preprogrammed within the genes that stay with the fetus and then with the adult throughout life. Somehow, they are cued to turn on only at this stage even though at this stage we scarcely detect any indication that a child is in the making.

The beauty and awe of life's genesis lies in the details. Skipping over the intricacies can lead a person to believe that milk just naturally comes in containers, and is not the end result of a process starting with sunlight shining on grass. What follows is just a sampling of the relevant steps in gestation, each revealing only a hint of the complexly ordered wisdom built into the process.

The morula, as the cluster of cells is called at this stage, floats freely in the uterus until seven or eight days after ovulation, absorbing its needed nutrients from uterine secretions. Internal timing now somehow tells the morula to change shape from a compact ball of sixty or so cells to a fluid-filled sphere, the "blastocyst," housing those previously starved cells destined to become the embryo. The outer cells of the blastocyst are to evolve into the fetus's part of the placenta, the semipermeable interface between mother and "child."

The difference in the genetic composition between the mother and the fetus makes the fetus appear as a foreign invader, and therefore subject to attack by the mother's immune system. The placenta, a product of both the fetus and the mother, has much of the task of isolating the fetus from the mother. In clever fashion, it allows transfer of maternal nutrients and oxygen to the fetus while expelling fetal wastes out to the mother's circulatory system. It's a lesson plan in chemical engineering.

Approximately seven days after fertilization the blastocyst decides to contact the wall of the uterus, release enzymes that literally digest (!) that part of the uterus, and in doing so, mesh with the uterine wall. The entire cell mass consists of a few hundred cells, and is barely larger than the original egg, still about the size of this sentence's period.

Now something quite surprising happens (as if the entire process till now had not also been surprising—but the next step is unique in life). Some of the outer cells divide and mesh together as if many individual cells decided to join forces and do away with the separating membranes. The resulting multicell structure projects capillaries further into the uterine wall, where they exude enzymes that dissolve the mother's capillary-sized arteries. (Don't lose sight of the fact that these enzymes had to be manufactured according to information, wisdom stored in the egg's DNA. Wisdom within wisdom.) The multi-cell structure has become the placenta. With the blood vessel walls ruptured, the mother's blood floods into the spaces between the fetal capillaries, which absorb the blood's nutrients and oxygen and transport those life-sustaining factors back to the blastocyst. Baby growth can now commence. A bit over one week has passed.

The seemingly dormant cells within the blastocyst awaken and by the end of the second week they align to form a longitudinal axis. Differentiation has begun. Some still unknown mechanism will turn on a set of genes in one cell and then turn it off, while orchestrating a totally separate set of genes in another cell or group of cells (all of which were identical a few days ago). The result is that the originally identical cells become very different parts of the body.

Until now, the cells held only the potential to become the various organs of a fully developed body. Now that potential comes into play, causing differentiation within what was originally a unity. And all according to an intricate ballet.

By two to three weeks, structures that will become the heart and nervous system can already be identified. A bit over a week later the heartbeat begins, though it is not yet palpable to the mother. By week five the key organs are in place: liver, pancreas, thyroid. Beginnings of eyes dot the sides of the head. The embryo is still a mere quarter inch long (a bit over half a centimeter), about the length of the "the" in this sentence, but as full of variety as an encyclopedia.

At the sixth week, precursor cells mark the beginning of the central nervous system. Eyes are now clearly visible. The brain begins to enlarge. It will continue to develop even long after birth. Arm and leg buds each sport five digits fully joined by webbing, somewhat like a duck's webbed feet. During the next two weeks, the cells of the web die, apparently so programmed in their genetic makings, and the fingers separate. Most of the crucial action is complete by the end of the first eight weeks—a transition from embryo to fetus having the beginnings of all the primary organs and limbs of the baby-to-be.

By the ninth week the arms and legs, hands and feet, and most of the body look clearly human. Eyes have begun to migrate from the sides of the head toward their human forward position.

I mention the forming of the central nervous system, eyes, an arm, a finger. The few words trivialize a world of complexity. Each organ is the result of a metropolis of activity.

Consider the eye. Look out the window. Your field of view catches a vista perhaps a mile wide. It all appears projected onto half of a sphere at the very back of your eye, the retina, less than three centimeters in diameter. Yet your brain sees within those three centimeters of information a world a mile wide and knows it is no Disney cartoon the size of a postage stamp. Light from the outside world has reached your retina with only slight distortion. That's because somehow those clever genes in your body produced crystal clear, transparent cells for the eyes' outer casing, the cornea and the lens just behind the cornea, and the thick fluid that fills the globe of the eye between the lens and the retina. Amazing. All those totally clear cells and fluid even though most of our body is opaque or translucent. Some cells of your eyes are yours for life. As you age, more are added, but the ones you were born with are still with you as well.

The iris, which is controlled by an array of muscles, regulates the amount of light entering based on feedback from the retina. Behind the retina is a heavily pigmented layer that absorbs light not captured by the retina. A second array of muscles changes the shape of the lens, bending the light more or less as per the extent of the lens's curvature, focusing the incoming images sharply on the retina. (All land vertebrates use this system to sharpen the image. A fish lens acts in a manner similar to a camera, focusing by moving the lens backward or forward.) Of course, the concept of focusing assumes the brain makes some decision as to what a "sharp" image means. Might the world really be blurry and we just see it as sharp?

All those muscles working in unison with no conscious thought on your part, and all in the blink of an eye, and all originally stored in one fertilized cell. But let us take one final tour-within-the-tour, to show the full miraculousness of this tiny datum within what was once a single cell. Because when the light reaches the retina, a three-ring circus of activity begins.

The light passes through the cornea, is bent and focused by the lens, passes through the vitreous humor, and ultimately arrives at the surface of the retina. But this surface is not sensitive to light! Still the eye can record even a single photon of light. First the light ray or rays must further pass by a stacked array of photomultiplier cells that vastly multiply the signal they receive from the retina. Finally, at the very back of the retina, which makes it the very back of the eye, are the light receptors, some hundred million rods for colorless sensitivity in dim light, and another approximately seven million cones for color vision in bright light. (Notice that in very dim light we do not see colors. The world becomes various shades of gray. That is because in reduced light, only the rods of our retinas function.) (Figure 6.)

We mentally build the range of colors our brain "sees" by mixing the inputs from three types of photoreceptors: sensitivity to low-energy red colors peaks at 630 nanometers (billionths of a meter) wave length; intermediate greenish peaks at 560 nm; and the high-energy end of the visible spectrum, blue, peaks at 420 nm. The visible range of light, 380 nm (deep purple) to 750 nm (dark red), is only a tiny fraction of the total range of electromagnetic radiation from radio waves at 100 meters in length and more to high-energy gamma rays measured in the thousandths of a nanometer.

Housed within the deepest section of the photoreceptors is a pancake-like stack of discs containing light-absorbing pigment specific for one of the three color/photon energy ranges. Ab-

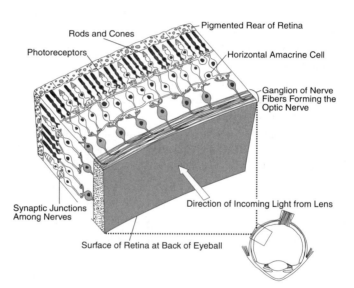

Figure 6

Detail of a Retina

Note that the photoreceptors are located at the very rear of the retinal layer and therefore at the very back of the eye. Light reaches these receptors only after it has maneuvered past the multitude of nerve fibers. If we had not learned in an earlier chapter that, according to the Bible, intelligent design, even at the level of the Divine, is not necessarily perfect design, we might erroneously find here an argument against the concept of a metaphysical presence within the physical world.

sorption of an incoming photon by a pigment molecule starts a chain reaction that heads back up the stack of nerves toward the surface of the retina. In essence, the human retina is designed inside out.

The immediate receptor of the photon is a derivative of vitamin A, termed retinol. A two-stage reaction then occurs that shuts down the receptor after its having fired. In very bright light, this reduces its sensitivity, thus adjusting the eye's signal to the brain. The photoreceptor, now energized by the absorbed photon of light, induces a cascade of as many as five hundred molecules of signal protein, each of which binds to and in doing so activates an enzyme that reacts with up to four thousand secondary molecules.

Sodium ion channels in the membrane, normally open in the absence of light, now close in response to the enzyme activity. This changes the electrical potential (the voltage difference) across the cell membrane and starts the nerve signal on its way to the brain, traveling up through the stacked chain of retina nerve cells that link rod or cone to the long ganglion nerve, the avenue to the brain.

Along the way, while traveling at 100 meters a second, the signal passes by a long, thin, laterally arranged nerve, the amacrine cell. This beauty of engineering compares inputs from adjacent receptor nerves to check for motion in the viewed image. Ever notice how a falling leaf or a moving hand barely visible out of the corner of your eye catches your attention? It's amacrine doing its job, protecting you from an end-run surprise attack. Frogs have the same array but to a far greater extreme than we humans. They respond only to moving objects. A stationary fly arouses zero interest in the frog's brain. If the fly moves, it soon becomes lunch.

Ganglion cells by the millions, leading from all the receptors, join together to form the optic nerve, and being at the forward, that is the lens side, of the retina, must now pass back through the retina, and head for the brain. This puncture through the retina by the optic nerve produces the famous blind spot we all have in our vision. We don't see the blank because our brain produces a conscious image that smears over the lacuna.

The multitude of ganglion cells cannot be treated like strands of cooked spaghetti tossed into a tube. The relative positioning of each receptor must be retained in the positioning of its ganglion so that the signals map onto the brain in accord with the image. That is quite a feat of bio-bookkeeping. But we manage to do it without even giving it a thought. The information these nerves bring to the brain represents points of light. Our brain blends those points into smooth, continuous images. If it didn't, our every view would appear as a Seurat painting, filled with dots of color and shade.

The menagerie of molecules that prepared the body for sight was entirely hidden by my simple words: eyes formed. Sometimes we see the trees and sometimes the forest. With life, there are forests even within the trees. Wisdom within wisdom.

BIRTH is no less complex.

The fetal placenta, "realizing" that the time for birthing has arrived, stops producing a hormone that for nine months kept the uterine muscles relaxed. Without this chemical inhibition, they begin the contractions of labor. As the actual birth begins, the umbilical arteries, for nine months the fetal life line, clamp shut while the umbilical vein amazingly waits to shut just long enough for the fetal blood that happened to be in the placenta at this moment to be returned to the fetus. Is this amazing, or not?! Blood flow is rerouted to pass through vessels of the lungs and intestines as the lungs prepare to breathe in air and the stomach prepares to digest food, the well-balanced nutrients of mother's milk. A hole between the two sides of the heart that had allowed the fetal flow pattern closes as the direction of the flow of blood through the heart reverses.

It's our metamorphosis. We've changed from a tail-bearing organism immersed in liquid with eyes on the sides of our heads, a yolk sac, bronchial arches, a one-chambered heart, a snout-like nose, with webbed hands and feet, and become human. If you could relate this to a butterfly, would it believe such a transition is possible?

It was all programmed into one cell just nine months earlier. Amid the phenomenal complexity of biology, there exists an underlying simplicity so eloquent that it is expressed in a single molecule. In life, oneness is what we represent. A single cell at fertilization contained within it all the potential that you were ever physically to become. And every cell within your body retains that wisdom.

The unifying cohesiveness we found in the laws of physics has come shining through in the workings of biology as well. But in both fields, we see it only when we seek it. A superficial reading of the world teaches of a reality built on entities as diverse and unrelated in their composition as the steel of a gun barrel and the fragrance of a rose. Can they actually have anything in common? Indeed they can and do. Surpassing the harmony of the laws of physics and the workings of biology lies the subtle truth that every bit of existence is composed of a single substrate—energy—created at the beginning. Einstein theorized and a multitude of experiments have since tested the truth of this bizarre and totally illogical reality.

The Bible relates, in the usual English translation, that "the earth was unformed and void" (Genesis 1:2). But the Hebrew of "unformed and void" (*tohu* and *bohu*) has a much deeper meaning. The kabala teaches that *tohu* is the solitary primordial substance, created at the beginning from absolute nothing, the substanceless substrate from which all that is material was to be formed; and that *bohu* is a composite word, as are many Hebrew words of the Bible, meaning *bo*—in it; *hu*—there is. "In it [in the *tohu*] there is [potential]." The single substance filled with full potential from which the world would be constructed was created at the beginning. The hidden potential within the single fertilized cell from which the entire structure of a human arises is a mirror in miniature of the creation.

Yet as remarkable as the underlying unity of the physical world may be, science is on the brink of discovering an even more sensational reality, one predicted almost three thousand years ago, that wisdom is the basis of all existence. "With the word of God the heavens were made" (Ps. 33:6). "With wisdom God founded the earth" (Prov. 3:19). "With wisdom God created the heavens and the earth" (Gen. 1:1). Wisdom is the building block, the substrate, from which all the time and space and matter of the universe were created. Wisdom is the interface between the physics of the world and the metaphysics of creation.

To search out the link between the two will take more than results from the laboratories of physics and molecular biology. Wisdom is not weighable, nor is it easily stained and seen under a microscope. It is going to take a bundle of abstract thought, before which it would be productive to understand how we think. We've reached the workings of the brain.

6

NERVES: NATURE'S

INFORMATION NETWORK

The essence of life is found in the processing of information. The wonder of life is the complexity to which that information gives rise. The paradox of life is the absence of any hint in nature, the physical world, as to the source of that information. As reluctant as I am as a scientist to admit it, the metaphysical may well provide the answer to this paradox.

Wrapped in smiles of love, I hug a giggling baby. A weight lifter exhales as he presses a three-hundred-pound barbell. A composer thinks through the notes of a symphony in the making, her brain on fire with melodies. The brilliant red of yesterday's sunrise emerges from my memory and once again I am filled with awe. All these experiences and a myriad of others that constitute our mental and physical lives rely on the near lightning-fast transmission of impulses within and among the labyrinth of nerves that lace our body—a million miles of bioelectric cables. If we are going to understand how our brains function and, more importantly, how the conscious mind emerges from the physical brain, we must first understand the functioning of nerves, for in essence the brain is a highly structured mass of nerves.

In simple terms, the nerve is the basic structure used in life for sending information from one body part to another. What is surprising is that the body uses the same means of transmission regardless of the type of information. Be it touch, taste or smell, sound or sight, the data related to these sensations or thoughts are encoded as pulses of electricity, voltage peaks, that travel as waves within cleverly insulated extensions of the nerves, known as axons. Upon reaching their targets, be they another nerve or a muscle fiber, the pulses of voltage release a chemical transmitter that passes to the target. The resulting chemical change in the target cell might cause a muscle to contract or a thought to arise. Within the neurological maze of the brain, these identical electrical signals are then categorized among the range of our varied senses.

In some cases the timing of the signals is exquisitely fine. In speech, for you to differentiate between the enunciation of a "b" and a "p," your lips must open some thirty thousandths of a second before you cause your vocal cords to vibrate for the "p" sound to emerge rather than a "b" sound, which occurs when you open your lips and vibrate your vocal cords simultaneously. Thirty thousandths of a second. Consider what this reveals concerning the precision inherent in mental and neurological processing. It's a sliver of time that makes the difference between bat and ball and Pat and Paul. Your brain determines this phenomenally tight timing and you don't even have to "think" about it. It's probably controlled by the brain part located close to the brain stem known as the cerebellum. The entire sequence is encrypted when the signal to vocalize a "p" or "b" arises in your thoughts.

Could this complex yet ordered precision have evolved without guidance? The problem in answering that question is not really knowing what Darwinian theory has to say about the process. While there is a clear description, in the theory and in the fossil record (and in the Bible too) of a development of life from the simple to the complex, there is no statement in any of these sources as to how the development of complexity came about.

A theory sets forth a concept of how specific events occur. It is not merely a description of those events. A careful look at the current theory of evolution reveals not a theory, but merely a description of the "punctuated" jumps in the fossil record. The evolution of life, if evolution is the proper word, is indeed punctuated. The world awaits a theory for what processes might have yielded those punctuations. It is for this reason that when the London Museum of Natural History, a bastion of Darwinian dogma, mounted a massive exhibit on evolution, occupying an entire wing of the second floor, the only examples it could show were pink daisies evolving into blue daisies, little dogs evolving into big dogs, a few dozen species of cichlid fish evolving into hundreds of species of—you guessed it—cichlid fish. They could not come up with a single major morphological change clearly recorded in the fossil record. I am not anti-evolution. And I am not pro-creation. What I am is pro-look-at-the-data-and-see-what-they-teach.

Darwin realized the problem when he advised us to use our imagination to fill in the punctuations. He insisted in the Introduction to his *Origin of Species* that despite the "steady misrepresentation [of his theory]," natural selection is "not the exclusive means of modification. . . ." In the closing lines of that famous book, Darwin elaborated on this idea. "There is a grandeur in this view of life, with its several powers having been breathed by the Creator into a few forms or into one. . . ." Darwin saw in the wonder of life the need for a ghost in the system, the powers breathed by the Creator. The laws of nature instilled in our universe and the physical conditions on the earth are ideal for sustaining life. But nowhere in these laws or conditions is there a clue for how the organized information of life originated or developed. The wisdom intrinsic to the simplest forms of life is nowhere presaged in the substrate from which life is built.

Don't let the grandeur of your body's origins and organization escape you. No, on further consideration, better not to think about it. You may become tongue-tied. Studying these brain/nerve/muscle interactions has all but tongue-tied my game of squash. I can't swing the racket without marveling at the train of bioelectronic signals that must be pulsing through my body's neurons.

I tense as my wife's shot slams off the front wall. Visual stimulation of eyes produces a chemical reaction in the retinas' photoreceptors, retinas' signals to optic nerves, optic nerves to brain, and brain to motor nerves that will pulse signals through their axons at over 100 meters per second to a phalanx of muscles controlling eyes, toes, ankles, legs, hips, shoulders, arms, wrist, fingers. They're all in on the act as I try to return Barbara's slam from just above the red line. Voltage-gated ion channels in cell membranes open and in doing so allow an avalanche of calcium ions to race through those openings, causing extended muscles to contract. An arm swings and with a bit of skill, and luck, the strings of my squash racket slam into the small black rubber ball.

The process of biological information transfer is a tale of awe. Consider just one aspect of this bodily train of events. How does the brain decide that the two-dimensional image projected onto the eyes' retinas represents a three-dimensional world? After all, the visual image is converted into an array of electrical stimuli, each of which is a one-dimensional pulse of voltage, in essence a single point. These one-dimensional signals, carried by the mass of fibers that make up the optic nerves, are what the brain receives, upside down and reversed just to make the progression a bit more complex. How does the brain know that the identical type of neural signal, having originated in my ears as a reaction to the sound waves produced by the black rubber ball pounding off the back wall, should be decoded as sound? From where does it get its smarts?

For a small number of unfortunate persons, the wiring has gotten crossed, resulting in sight that is heard and sound that is seen. Ever get punched in the eye and you see stars? Why stars, why not just pain? You saw stars because the punch activated nerve endings in the retina, which, as good soldiers, sent an electrical signal to the brain. But to which part of the brain? The part that interprets an electrical pulse from those nerves as sight, not as pressure and pain. Hence the seeing of stars.

Never ever take the simple beauty of sentience for granted. It didn't have to be that way. Complexity underlies even seemingly simple acts.

The technical information on the structure and function of nerves that I am about to present is not intended to encourage you, the reader, to pursue a career in molecular biology or physiology. But it is intended to make you aware that when the alarm clock rings in the morning, and blurry-eyed, you fumble for the button to silence it, there's already in your sleepy body a symphony of cellular reactions in progress, which if you could put them to music, would by comparison make Beethoven's violin concerto seem little more than chaotic street noise. I urge you, skeptic or believer, to acknowledge this astoundingly integrated chain of events that was required to bring this symphony into being. The concept that life may have resulted from inert, dumb, random reactions starting in an undifferentiated ball of energy at the big bang stretches the imagination. Does this prove there's a God active in our world? I personally do not think that the complexity of life proves the existence of the Divine. But it does demonstrate unequivocally that we are missing some basic factors in how the origin and the development of life occurred.

Whether those factors include the metaphysical, or even the Divine, may never be absolutely verifiable. However, the final leap of faith for or against the concept that the metaphysical is active within the physical universe that it created (for both the skeptic and the believer require a final leap of faith) is best made from a position of knowledge. Faith backed by knowledge is much stronger than faith based on an emotionally driven gossamer hope, whether that faith be secular or religious.

From energy to rocks and water, to life, to consciousness. Now that's an impressive and even possibly Divine chain. Let's look at some of the biology that developed from the big bang.

In his book *How the Mind Works*, Steven Pinker decides "not [to] say much about neurons, hormones, and neurotransmitters [those complex molecules produced in nerve cells that allow communication among nerves]." Understand that, according to the title, Pinker's book is supposedly about what goes on inside our heads, that is to say, the output of interactions among neurons, hormones, and neurotransmitters.* Only by avoiding the intricacy of how that output comes about and what limits it is Pinker able to enthuse repeatedly throughout the book about the assumed but untested wisdom that we are purely "the product of natural selection." The complexity of stimulating and transmitting a single neural signal, let alone the functioning of all the other organelles that contribute to the finely orchestrated symphony we call life, belies the assumption. Let's mentally crunch through some of that detail.

In overall functioning, nerve cells are similar to all body cells. A sacklike membrane, consisting of two back-to-back layers of molecules, defines the boundaries of the cell. The water-loving (hydrophilic) heads of these molecules are exposed on the interior and exterior surfaces of the membrane. The central part of the membrane houses the hydrophobic (water-avoiding) legs of the membrane molecules. Channels, which open and close in

---

*A far better book on the brain/mind interface with much less ego pushing through and far more objectively written is, in my opinion, Rita Carter's *Mapping the Mind*. Both books are 100 percent secular in their approach. Notwithstanding their titles, neither book finds a mind within the brain. Pinker opts that since our minds are limited, we may be unable to solve the enigma of how sentience arises from the brain.

response to chemical and voltage signals from within and without the cell, provide controlled entrance and egress for molecules active in cellular metabolism. A nucleus within the cell concentrates the primary genetic material, the DNA. Mitochondria throughout the cell help convert nutrient glucose into the energy-rich molecule, ATP. Microtubules extend to all regions of the cell, providing structure and, of crucial importance, tracks along which motor proteins can transport needed molecular materials to reaction sites.

Two regions differentiate nerve cells from other cell types: axons and dendrites. The axon is the transmitter. Some 20 microns in diameter and reaching as long as a meter, the axon is the extension of the cell that carries the nerve's message as a bioelectrical signal, known as an action potential, to the target organ. The end of an axon may divide into a bush of terminals, allowing it to stimulate whole groups of targets simultaneously. This plays a central role when signaling a multitude of muscle fibers to come into action.

The dendrite portion of a nerve is the receiver of the axon's message. Fingerlike extensions of the dendrite increase its surface area, allowing it to receive multiple signals simultaneously. In some cases as many as one hundred thousand axon terminals reach a single dendrite. The nerve will sum these individual signals and decide if the total input message is strong enough, that is, exceeds a threshold voltage, to warrant the passing of this information on to another nerve. "The nerve will sum the signals and decide"—it sounds almost mental, almost human, yet it is only a batch of carefully arranged molecules (see Figure 7).

In a sensory nerve, such as those of a fingertip that feel touch, dendritelike extensions respond to stimulation of the skin and send a pulse along that nerve's axon toward the central nervous system, be it brain or spinal cord. Motor nerves then transmit signals received from the central nervous system to the muscles being called into action.

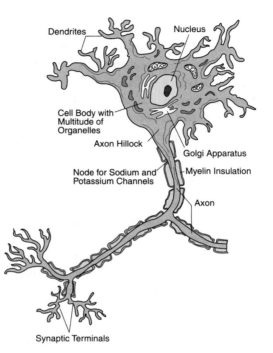

Dendrites

Nucleus

Cell Body with
Multitude of
Organelles

Axon Hillock

Golgi Apparatus

Node for Sodium and
Potassium Channels

Myelin Insulation

Axon

Synaptic Terminals

Figure 7

An Idealized Nerve Cell

Synaptic terminals by the thousands, coming from other nerves, cover the nerve cell's dendrites, each bringing its own electrochemical output. The resulting signal is converted into the bioelectric action potential at the axon hillock. The action potential then races down the axon to the synaptic terminals, where waiting chemical neurotransmitters will be released, sending the signal on to other nerves or muscle fibers. Axon length can reach a meter when transmitting signals from the tip of a finger or toe to the central nervous system.

Signal transmission by a neuron is totally dependent upon differences in voltage along the axon and across the neuron cell membrane. Our every thought and deed at some point reduces to a bioelectric signal. The voltage potential is induced by differences in concentrations primarily of sodium, potassium, and chlorine ions. (An ion is an atom with one or more of its electrons missing and so has a net positive electrical charge.) I'll focus here on sodium and potassium.

The concentration of sodium ions inside the cell is approximately ten times less than its concentration in the extracellular fluid just outside the cell membrane. The pattern for potassium is the opposite, being ten times more concentrated within the cell relative to the extracellular fluid. These inequalities are maintained by pumps that operate across the cell's membrane. We house an alarmingly sophisticated technology of molecular dimensions. The differences in ion concentrations produce a negative voltage within each cell of about minus 70 millivolts. Every cell in your body stands ready to give off a biological spark.

Large amounts of sodium can enter the cell only if the cell receives a signal to open voltage-gated channels that span the cell membrane. And that is the key to all neural signal transmission. Upon stimulation, these channels open and sodium ions flood into the cell. Via a positive feedback loop, the initial sudden increase in sodium stimulates more sodium channels to open. In a few thousandths of a second, an ionic avalanche of sodium raises the cell's interior from the original minus 70 millivolts to a potential of plus 50 millivolts. At this voltage, the cell is poised to fire.

Assume a nerve's dendrites have picked up a signal sufficiently strong to warrant its propagation to a further nerve or muscle. The neuron is primed for response. Try to imagine in this discussion how the synchronized, intertwined complexity about to be described could have evolved from inert and unthinking rocks and water, for that is what preceded life on the originally sterile earth.

An adjacent axon has signaled to the neuron that it is to generate a signal, the bioelectric action potential. Perhaps you've touched something too hot for comfort. The heat stimulates the sensitive endings of the nerve, inducing it to send the message to its target receivers, in this case rapidly to the spinal cord and slightly less rapidly to the brain.

The signal, a cascade of ions, travels from the receiving dendrite, past the cell body, and on toward the axonlike extension. At this point the action potential is generated that, as a wave, transmits the signal the length of the axon to the synaptic terminals at its end. Since the axon terminal does not actually attach to the target neuron's dendrite, nature had to invent a method of getting the message across the gap measuring approximately 20 billionths of a meter (20 nanometers) that separates axon terminal from target dendrite. Nature was up to the task. The nerve accomplishes it by having the electrical action potential within the axon stimulate the release of chemical neurotransmitters into the synaptic gap. The electrical signal has become a chemical signal. These neurotransmitters have been "conveniently" stored in organelles called Golgi apparatus near the axon's synaptic terminals. The Golgi are budlike globular beauties of nature that package the neurotransmitters at their point of manufacture in the cell body and then, with the help of motor proteins, transport them and other essential molecules from within the cell body, down the axon, to the location of use near the cell membrane. The Golgi, upon command, release the neurotransmitter into the synapse, where it diffuses across the opening, attaches to the target dendrite, and in doing so triggers a secondary neural signal to start on its way.

Consider the implications of just one aspect of this event. Somehow the Golgi apparatus was positioned at the terminal by nature. Way back at the cell body, near the start of the axon—which in some cases is a meter distant from the terminal—a signal had been given to the DNA to provide the pattern to make messenger RNA, mRNA, which, with the help of transfer RNA, tRNA, and a few other cellular microorganelles, would churn out copies of the one specific type of a wide range of potential neurotransmitters needed for this one type of stimulus, package it in Golgi buds, and via motor proteins that literally walk molecular step by molecular step along microtubule tracks the length of the axon, carry the Golgi loaded with neurotransmitters to the axon terminal area, there to wait patiently in the wings until called upon by the axon's electrical signal—the action potential—to move into action. The trip from the cell body where the Golgi and neurotransmitter are made to axon terminal takes about two days when traveling via motor protein. If nature had relied on diffusion to make the trip, the journey would have taken about two years. When called into action, the Golgi move within a millisecond.

The Golgi bud fuses with the inner surface of the axon synaptic membrane, and then, in a process known as exocytosis, bursts through on the outside, into the 20-nanometer-wide synaptic gap. The electric signal of the axon has been converted into the chemical signal of the neurotransmitter.

Golgi buds might be likened to a vitamin pill. They concentrate the one type of needed molecules and then release them as a batch. For neurotransmitters, Golgi concentrate them at two hundred times greater than would be possible if these molecules were just free floating at the synapse terminal. It's all so clever. None of this wisdom is even hinted at when we view life from the outside. The insights of molecular biology have revealed a complexity at every stage of life's processes such that, if we were forced to rely on random mutations to produce them step by step, in the words of Nobel laureate de Duve, "eternity would not suffice."

The huge concentrated input of neurotransmitters released into the synapse causes them to diffuse rapidly across the synaptic gap. In less than a millisecond they reach the dendrite surface. If the neurotransmitter is the correct one for the job, its shape will complement the shape of a receptor on the dendrite membrane and it will bond. The right key in the right lock—only one fits and nature designed it just so. It is either the result of chance random reactions among rocks and water or the expression of an underlying wisdom poking its head through into the physics of life. Those are the only two choices available.

The amount of released neurotransmitter contains a large excess over the quantity actually needed to bind with the target dendrite. The excess remaining in the synapse blocks the input nerve from transmitting an additional signal. If nature were to rely on diffusion to have the excess molecules drift away, the down time of that nerve would be considerable, measured in seconds and even minutes. But nature foresaw the problem and is neither so lazy nor so patient. In time spans measured in a few thousandths of a second, the excess is absorbed by the axon terminal and the remainder decomposed by enzymes, catalytic proteins fashioned way back in the cell body from DNA-held information that is transcribed via mRNA via ribosomes and tRNA, and so on and so on, and then released into the extracellular plasma to be on hand just in case it is needed to reduce the complex neurotransmitter molecule to its component parts.

Of course the lock-and-key system of the neurotransmitter/dendrite interaction is exactly what a Darwinian-type survival-of-the-fit would favor as a fail-safe system to block out wrongly directed signals. Any survival system would favor it. The question is how did the system arise? In the late 1970s, a symposium was held at the Wistar Institute of Anatomy and Physiology in which mathematicians forced biologists to confront the reality that all calculations of probability say no to the assumption of randomness being the driving force behind life's development. But cognitive dissonance held sway then and still holds today. The conclusion of the biologists was, and remains, that the mathematical assumptions must be incorrect since evolution must have occurred. As I discussed at length in *The Science of God*, both the Bible and the findings of paleontology indicate that life developed from the simple to the complex. That development is not the problem. What is at question are the mechanisms behind the development. The calculations of probability haven't changed since the Wistar symposium, but the understanding of molecular biology has, and that understanding has revealed biosystems far more complex then those that were imagined in the 1970s.

The fastest of mammalian nerve cells can fire a thousand times per second. That requires depolarization of the cell membrane and the restoring of the ionic concentrations of potassium and sodium in less than a thousandth of a second. The ion pumps (molecules that actually pump) that do the job are a wonder of efficiency. Meanwhile the action potential is traveling in excess of 100 meters per second toward its target organ. But we are not dealing here with a copper wire where the electric current is facilitated by readily exchanged electrons within the copper metal. In the biosystem of an axon, the influx of bulky sodium ions through gated channels that pierce the cell membrane must propagate the signal.

The velocity of signal propagation depends upon two factors: axon diameter and how far ahead of the actual signal the sodium channels can be induced to open. The larger the diameter, the more rapid the propagation. To gain this advantage in signal transmission, some animals, such as the squid, have axons a millimeter in diameter, easily visible to the unaided eye. Conduction here reaches 25 meters per second. But axons of this size become prohibitively cumbersome if many nerves traverse the same region, as with the limbs of vertebrates. If this solution had been adopted by human nerves, the diameter of our arms would be measured in meters.

Vertebrates have developed a far more effective, and quite clever, method of increasing the velocity of impulse propagation. Nerves that must stimulate organs requiring rapid response—for example, those that might signal the need to remove a finger from a hot surface, or power a leg muscle to run from a hungry lion—are electrically insulated by a fatty molecular layer, a nonconducting lipid sheath known as myelin. This insulation greatly reduces current leakage (sodium loss) from the axon outward across the axon membrane. It also regulates the positioning of the sodium channels in the membrane.

A typical vertebrate axon is 10 to 20 microns (millionths of a meter) in diameter, fifty times narrower than the squid's nerve, and yet the nerve pulse in the human moves at 100 meters per second, four times faster than in the squid. How is this accomplished? The wisdom of biology doesn't try to beat the rules of nature. It outsmarts them. The secret is in the positioning of the myelin sheathing. Approximately every millimeter there is a break in the sheath, a node, a few microns wide. The sodium and potassium channels are concentrated at these nodes. As the impulse propagates along the axon, it draws sodium always from the nodes just ahead of it and, in essence, leaps forward to that node instantly. The result is signal transmission from the central nervous system spinal cord to a finger or toe in a hundredth of a second.

The action potential (AP), as it moves along an axon, represents a high voltage peak relative to the voltage potentials of the surrounding portions, both before and after, of that axon. In theory that voltage peak should be able to travel in both directions, both down the axon and back up toward the cell body, since in both directions, forward and back, the voltage is lower than the peak. To obviate the possibility of the AP reversing its direction and traveling back toward the cell body, nature has brilliantly invented a double molecular lock on the sodium channels. Once a channel has opened for the sodium cascade into the cell and then closed, that particular channel cannot reopen until the local region has completely depolarized. By that time, the AP has moved out of range, far along the axon toward the nerve's synaptic terminals. The molecular keepers of these gates are very clever. Somehow they have learned their lessons well.

The benefits of myelin sheathing are sadly most apparent in their absence. Multiple sclerosis is a disorder in which the autoimmune system in error destroys the myelin. The result opens the cell membrane to sodium loss across the exposed axon membrane. As the disease progresses and myelin depletion increases, transmission rates slow to a few meters per second. Eventually leakage is so great that the axon can no longer transmit its message. The target muscle becomes paralyzed. The fact that for most of us life's mechanisms work in proper order is a wondrous miracle. When they do not it is a tragedy.

The system described and diagrammed above is an ingenious one for communicating massive amounts of complex information. The parallel processing and perfect timing involved are as elegant as the finest supercomputer. Perhaps some day, in the age of communications technology now upon us, we will imitate and exploit our own design: In the meantime we can only wonder at the workings of our chemistry.

With this understanding of how nerves, the information-bearing cells of the brain, function, we are ready to enter the brain itself. Knowing the complexity of the processes involved, when we see a diagram showing how simple evolution is, how one organ can change into another merely by adding a feature here and there, we must realize that those demonstrations are a farce. As long as the intricate workings of the cell are disregarded, there's no problem for a Steven Pinker, or Stephen Jay Gould, or Richard Dawkins to talk of random reactions producing the goods of life.

It is hard not to be fooled by the foolish arguments when they originate from intelligent foolers. Abraham Lincoln is quoted as having said that while you can fool some of the people all of the time, and all of the people some of the time, you cannot fool all of the people all of the time. The more knowledge one has, the harder it becomes to be fooled.

Those diagrams that in ten steps evolve from a random spread of lines into people-like outlines, and in a few hundred steps simulate a light-sensitive patch on skin evolving into an eye, once had me fooled. They are so impressively convincing. Then I studied molecular biology.

7

THE BRAIN

BEHIND THE MIND

If the universe is indeed the expression of an idea, the brain may be the sole antenna with circuitry tuned to pick up the signal of that idea.

When I think a thought, what is understanding what? Just to ask the question I must conjure up internal images. When I think of myself, I recall an image which, if I saw it in a photo, I'd recognize as myself. Actually the immediate image that surfaces is me in my roaring twenties. That's my psyche's image of self. I have to work logic into the picture if I am going to see today's wrinkles and receding hairline.

Surgeons studying activity in various parts of the brain have discovered that as they stimulate regions of the external body, from head to toe, neurological reactions in the brain produce coherent maps of the body, albeit upside down and left–right reversed. Some portions occupy a greater cerebral area than they actually do on the body. The mental maps are, in a sense, reminiscent of a cover that appeared on *The New Yorker* magazine several years ago. It depicted the artist's concept of a New Yorker's view of the world looking west. Manhattan and New Jersey fill the entire foreground, followed by a very thin sweep of the Midwest. Then Los Angeles and San Francisco loom, bordered by a strip of water, the Pacific reduced to a trickle. Hawaii and China fill the horizon. The body, as the brain sees it, also follows the correct sequence of limbs from feet to head, but the emphasis reads like a caricature of the human body—which of course it is. Our psyche is a caricature. Our very, very big feet all but abut our oversized genitals. Legs, having fewer neural receptors, get short shift in the cranial map. Like the Midwest to the New Yorker, the legs are there merely to connect feet to groin. Arms get a bit more space. Hands are huge. The face, especially lips and tongue, are similarly enlarged.

When I think of specific body parts without even physically touching them, the corresponding neurons on the mental map fire. Science fiction and science fact use this thought-to-neuron connection to envision mental control of the world around us. At the neural level, we are actually witnessing a continual display of mind over matter. Or possibly more accurately stated, the consciousness of the mind over the consciousness of matter.

In earlier chapters we've seen that the universe exhibits the essence of a mind, a wisdom behind and even within matter. This universal wisdom, sometimes defined as God's presence, is the essence of the metaphysical as it projects into our finite, physical world. If, in fact, humans are created in God's image, as the Bible claims, the task of this and the coming chapters is to determine if mind—consciousness—precedes matter in man. This will be no small task in a culture steeped in what George Gilder refers to as "the materialist superstition," a worldview in which emotions, mind, and all feelings of spirituality are the products of the physical body. The religion of materialism quintessentially believes that if you can't measure it, weigh it, stain it, and see it under a microscope, it does not exist. No one has yet managed to weigh a mind.

BY birth, many neural connections are already hardwired, set in place by the genetic instructions of the cell. But many are not. We have something in our brains like the fixed read-only memory on a hard disk and then a self-correcting software program that goes along with the ROM. The software of the brain is malleable, reacting and developing in response to ongoing environmental stimulation. Cover the eye of a kitten or allow a childhood cataract to go uncorrected and that part of the brain's body map withers as neurologically adjacent organs take over the brain space originally dedicated to those impaired body parts. As long as there is life in the body, the brain is ever changing in response to that which life brings to the body. Stimulation and love are the recipes the young brain calls for, notwithstanding recent claims that infants can get along without them. Neglected toddlers do survive, but the neural physiology of the deprivation is engraved within their brains.

From all we know concerning the physiology of the brain, the sensory inputs, motor outputs, and interacting neurons, can we tease out the seat of consciousness, that which the mind experiences as self? Whether or not the mind has an aspect that transcends the physical brain is currently a moot point. At the minimum, awareness of what we feel as the mind appears to derive from activity in the brain. This amazing organ of perception, once by necessity treated as a black box—sensory spur in, motor and emotional response out, with no clue as to the intervening processes that linked input to output—is now open to investigation. Revelation of the previously unimagined intricate workings of the brain has challenged the simplistic theory of life's random evolution in a manner similar to the challenge presented by the discoveries of the complexity of cellular molecular biology. The difficulty in displacing the belief in Darwinian evolution, even though the theory fails to describe reality, is that no other materialist mechanism can explain the development of life as displayed by the fossil record and as described in the Bible. In a world so steeped in the physicality of materialism, calling upon metaphysical solutions is out of bounds.

With the help of a range of recently developed instruments,

we can at last begin to define boundaries to the processes of the brain, isolating and identifying those aspects of consciousness that are certainly the products of neural activity, homing in on the field of what might be a mind distinct from the flesh and blood of the brain. All the topics discussed in the foregoing chapters, the detailed structures and functions of cells, the biomechanics of information transmission in nerves, come together in the brain.

The adult human brain has approximately one hundred billion neurons (nerves). So does an infant's at birth, though many of the connections among the nerves have not yet formed. One hundred billion is also the approximate number of stars in our galaxy, the Milky Way, and the estimated number of galaxies in the entire visible universe. In an adult brain, the axon of each neuron connects with as many as a hundred thousand dendrites of other neurons. The branching is stupendous, a million billion connections. That's 1,000,000,000,000,000 points within our heads at which neurotransmitters are racing, sending information from nerve to target nerve. A massive web of activity contained within just under one and a half liters of volume, with most of our conscious thought emanating from a layer 2 to 4 millimeters (about one-eighth inch) thick, the cerebral cortex, which covers the very top of the brain.

Here's a brain test: Wiggle your toes on both feet, and now at the same time wiggle your fingers, extend your arms and move them in horizontal circles with none of these motions being in the same direction, and while doing all this, gently shake your head right to left, left to right and quietly whistle a tune or recite a verse. And now consider all the neurological and molecular complexity involved in every one of those muscular contractions, all of it being induced by a flood of signals processed simultaneously in the central nervous system that started with signals from your optic nerves as you saw and read these words. For each individual motion and thought, the signals (action potentials) travel the length of axons of motor nerves extending from brain to muscle; sodium channels in axon membranes open and close; sodium and potassium pumps restore the electrolyte balance across each cell membrane so the same nerve can fire tens of times each second; neurotransmitters transported by the Golgi apparatus are released into synapses at the axon terminals and bond to adjoining dendrites. Calcium channels in the muscle fibers open to admit calcium ions that bond with proteins that bend and pull back other proteins, contracting a muscle fiber just enough so that the combined action of a million links each cycling five times each second produces the desired smooth force as you twirl your arms, move your head, whistle your tune, move your fingers and toes.

Of course you could never possibly do this on a regular basis and certainly not without concerted and conscious effort. But the wonder that your body represents, driven by a wisdom hidden within every particle of the universe, does the equivalent of it at almost every waking moment of the day. You start to cross a road and turn your head to check for traffic, step out with just the right gait from your leg muscles, judge the distance and timing of an oncoming car. You then turn to look for traffic in the second direction, hear a familiar voice (familiar, meaning recalled from memory), link that voice to a battery of other locations in your brain, the face, personality, and name that go with the voice, feel positive emotions toward the person, call out her or his name adjusting your vocal cord tension and lip shape to the task (without a thought, conscious at least, that it is only a difference of 30 milliseconds in lip/vocal cord timing between "P" for Paul and "B" for Bill), wave hello and yet somehow make it safely across the street, shaking hands with just the proper firmness of grip even though a few moments before you were pressing two-hundred-pound free weights at the gym.

It's called parallel processing, multiple tasking. The brain does it by the millions every waking second of every day, without even thinking—or better said, with barely a conscious thought. I wonder which is more of me: the conscious or the subconscious me? All our actions are preceded by thought, whether conscious or not. We enjoy free will. I cross the street only if I choose to. But is it my mind or is it my brain that chooses? Are they different? An overview of the brain's anatomy and function provides part of the answer.

Let's look at the development of those brainy organs that just orchestrated our walk across the street.

The first recognizable body part in a human embryo is the central nervous system (CNS). Eventually it will become the brain and spinal cord. Its early arrival is not surprising considering the fundamental role the CNS will play throughout life. By just two and a half weeks after fertilization a hollow trough has formed along the embryonic axis as cells migrate in from its periphery. Less than half a week earlier the first longitudinal organization of the embryo's cells had begun. By three weeks, the trough closes to form the neural tube. The entire embryo is still less than two millimeters (about a sixteenth of an inch) long.

During the third or fourth week the heart begins to beat, but not via stimulation from the brain or CNS, though surprisingly it is located near the brain in what will become the head. The heart is still one chambered, like that of a worm. In time the embryo and the heart will fold, moving the heart to the chest and forming the four-chambered heart of an adult human. For now, held within an embryo not quite three millimeters long, it has another two to three billion beats ahead of it, pumping life-giving nutrients to all parts of the body.

Within another day or so, two swellings, like small inverted cups, begin to protrude from the brain, the beginnings of the eyes. Eyes are, in a sense, externally visible extensions of the brain.

By thirty-five days old, the cerebral cortex, the part we associate with conscious thought and intelligence in an adult, has become visible. It's the part of the brain we are referring to when we say someone has a lot of gray matter. Gray is the color of cell bodies and dendrites, the nerve parts assumed to be allied with information storage. The brain starts to enlarge, just the beginning of a process that continues for years. Because the mother's pelvis must support the entire weight of her upper body, it requires considerable bone mass. This limits the size of the opening through which the fetus must pass at birth. The head at birth literally stretches the envelope. The brain at birth is a quarter of the mass of an adult's. Four-legged animals have relatively larger birthing channels since their pelvis need only support half the body weight.

So why, you might ask, are we humans, with our large brains and distinctive abilities, bipedal? Is this a flaw in the wisdom of design? Could we be smarter and mentally richer beings if we walked on all fours and thus had larger brains at birth? Not likely, as we see from the realm of the four-legged world. The wisdom in our design is that higher intelligence requires hands freed from supporting the body. Without that freedom to form complex tools, control fire, develop technology, there'd be no point to a larger brain. We are as large and as small as we need be.

In a way, the brain has outwitted the size restriction. The cerebral cortex and neocortex are wrinkled in a way remarkably similar to the shell of a walnut. This wrinkling has increased the cortex area without needing to expand the total brain volume, and therefore the size of the head. Once again wisdom has bypassed nature's constraints.

The brain produces neurons at an astounding rate in the womb. By birth there are one hundred billion. During the nine months of gestation, that averages out to between four and five thousand new nerves each second. Four to five thousand phenomenally complex axon-elongated cells constructed each second, each one racing out to find its target organ. These are guided on their journey by a trillion (a thousand billion!) structural glial cells. Don't just gloss over these fantastically huge numbers. Each cell houses all the complexity we discovered earlier: nucleus, DNA, mRNA, tRNA, ribosomes, motor proteins, ion channels, and on and on. And all are being manufactured at the rate of five thousand per second. The miraculous nature of life is found in its details.

By eight weeks, the embryo, with all its major body parts now visible, has become a fetus. At this time a very politically incorrect phenomenon occurs in the male, XY genotype fetus. Pulses of testosterone, associated with testes formation, are produced, altering the brain's development relative to the development in the XX female genotype. I dread to write this at the risk of being labeled sexist, but here goes. In general, the difference between XX and XY brain development appears to make girls better in speech and social relationships, boys more advanced in spatial relationships. Not for every human, but in general. Neurophysiologists may have discovered what parents have sensed all along!

At about twenty weeks, 140 days of age, neural connections form between the cerebral cortex and deeper brain parts. Though the difference between the human genome and that of a chimp is estimated to be less than 1 percent, our cerebral cortex has ten times more neurons. Another five weeks and parts of the limbic system, the seat of stored emotions, link with the cortex. This union in adults allows deliberate control of emotions that might otherwise result in automatic explosive response, fight or flight, each time we are stressed.

While I was at M.I.T., two cars got into a tangle on the southeast expressway leading into Boston. The drivers started to argue and one whipped out a handgun, shot and killed the other, and then sped off. His limbic-cortex linkage was in dire need of reinforcement, though I doubt that such a plea in court would have gotten a sympathetic response from the judge.

Myelin sheathing of the axons in the frontal lobes is completed only after reaching full adulthood. Since the frontal lobes are where deliberate throttling of emotion can occur, and since they are not yet fully functional until the sheathing is completed, it is not surprising that teenagers are so often more impulsive than their parents.

We've seen how the brain develops, but that still does not tell us how our thoughts arise. How the organs of a mature human brain interact may provide a clue to the processes of thought. The details you, the reader, are about to endure are admittedly complex. But they are also miraculous wonders in their complexity. To understand what happens when the brain does its work is to understand the beginning of the mind.

Since the only part of the brain visible from the outside is the organ devoted to vision, the eye, that is a reasonable starting point. In a previous chapter we saw how an electromagnetic signal (light) gets from the eye's retina via the optic nerve to the brain. The processing of the signal, breaking it into its component parts, starts immediately at the retina, where some neurons record aspects of color, some motion, some boundary contrast. That is just the beginning of the mental deconstruction of the information.

When thinking about sight, think of the symphony of molecular reactions of each optic nerve as the multitude of nerves communicate impulses, analyzing each impulse, deciding whether or not to pass on the pulse to other regions and other nerves. Think wonder. And ponder how a batch of carbon, nitrogen, oxygen, hydrogen, and a few other elements got together to cooperate so very wisely, thousands, even millions of times every second, throughout the brain and body. Had Darwin known of the wisdom hidden within life, I have confidence that he would have proposed a very different theory.

A scene catches your attention. Perhaps you are watching a

squash game and the black rubber ball has rebounded from the far wall just above the red line. A good shot, your brain tells you. The electromagnetic radiation (light) illuminating the scene bounces off the ball and wall and red line, travels to each of your eyes, which move in unison due to a brilliantly but subconsciously synchronized set of six muscles controlling the motion of each eye. Though they move together, the eyes do not follow identical tracks. The fact that the eyes are separated demands that they move at different angles. This single aspect of brain/muscle linkage is an exquisite demonstration of complexity. If it were absent, we'd have double vision. The difference in the angles of the eyes helps the brain estimate distance between the eye and the viewed object. As the light enters the eyes, the iris adjusts the aperture to regulate the amount of light reaching the lens. The double convex lenses change curvature to focus the image, now inverted, on each retina, where the photons of light are absorbed by photoreceptors. The energy of the incident light rays produces an array of electrochemical pulses, sending information of motion, color, shape, and borders via the optic nerve to the brain. At this stage the entire scene might be reproduced, mapped directly on the cerebral neurons, and then recalled, but life is not as simple as it meets the eye. We don't actually see any of this.

Once the scene has been parceled at the retina into neural signals, the scene is never reconstructed into the light patterns that reached your eye. What you perceive as a fragrant red rose or a black squash ball is, in your brain, a myriad of nearly identical electrical signals devoid of color, scent, or shape, with each signal directed to the specific portion of the brain devoted to the particular aspect of the scene.

Here lies one of the fundamental differences between a mind and a computer. In storing data, a computer, using semiconductors, stores an image as an array of electrical pluses or minuses in voltage, binary data. This is not so very different from the nerve's axon/dendrite storage system. But there is an essential difference in the retrieval of the information. Based on that array of binary data, and emanating directly from it, electric circuits excite pixels on the computer screen that reproduce the original scene.

In the brain, there is no evidence of a "screen" on which, or within which, the original image is displayed for mental viewing. There is no hint in your brain as to how you see the words you are now reading. We know how the inputs of the scene or smell or sound are mapped in the organs of the brain. It's the replay that confounds us.

As the image of the scene passes through the double convex lens common to all vertebrates, the image inverts. What is on the left in the outside world reaches the right side of the retina; the right reaches the left, the top the bottom. Something like heels over head. (Did you ever notice how nonsensical the expression "being head over heels in love" is? Friend, if you are not usually head over heels, you have a very strange way of walking!)

In point-by-point transmission from each retina, a ganglion of a million or so nerves carries the image aft into the brain, where the two optic nerves—one from each eye—meet and divide. All ganglion cells carrying information from the right fields of the retinas of both eyes continue to the right side of the brain. Those from the left retinal fields of both eyes go to the left. This means that approximately half of the retinal input from each eye goes to the corresponding side of the brain. The mental image is still reversed in the brain, since that inversion occurred at the lens, prior to the retina. By comparing the images, now in stereo, the brain estimates distance and texture (see Figure 8).

The first stop on the optic trail is the thalamus relay stations, one on the right half of the brain and one on the left. We've moved a bit over halfway toward the back of the head. At this

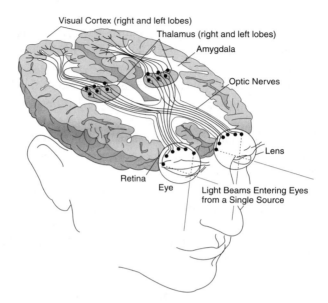

Figure 8

Visual Pathways Within the Brain

Note that the signals from the right sides of the retinas of both eyes go to the right hemisphere of the brain. Signals from the left sides of the retinas go to the left brain hemisphere. The amygdala, being closer to the eyes than the visual cortex, receives the signal before the visual cortex and can react to that information even though the amygdala produces no conscious visual awareness.

point, the signal is routed to the very back of the brain, the location of the visual cortex, and also to a nearby organ known as the amygdala. The pictorial part of vision is processed at the very posterior part of the brain, even though it begins at the most forward portion, the eyes. Emotional aspects of a scene are processed in the amygdala.

All incoming senses, other than smell, are inverted the way sight is and all have the thalamus as the first station. The neural circuitry of the thalamus tackles an intense amount of book-keeping since all the inputs are identical electrochemical pulses. Best not to mix auditory with retinal signals, or you'll be hearing sight and seeing sound. Smell, a most basic of senses, by-passes the thalamus and is routed directly to the amygdala.

It is the amygdala, one of a group of organs at times collectively referred to as the limbic system, that evokes emotional memory even when at times that memory evades our logic. During a recent fall semester, a student confided to me that he wasn't able to concentrate while sitting next to a particular member of the class. He could not understand why. A bit of questioning revealed that the aftershave lotion used by the other man was the one he had favored during a previous summer's escapade. It evoked in him all sorts of bittersweet feelings he had been unable to explain.

While all vertebrates have the emotion-packed limbic system, only mammals have a highly developed cerebral cortex, the site of advanced logic and data processing. The wisdom of the cortex notwithstanding, the amygdala, because of its proximity to the thalamus, induces its response before the cortex can get into the act. The limbic system is the fast route. The cortex is still working on the data when the amygdala is screaming for action. From here arise the responses of fight and flight, and, interestingly, a third emotional choice not usually mentioned, submission. Depending upon one's past experiences, the amygdala quickly chooses a response to incoming sensory information and prompts a response. The first part of us that acts, our first decision maker, is also our least logical self.

Even with the "slow" processing of the cortex, we're talking about milliseconds. Slow is still quite fast. But the difference in timing is unfortunately sufficient to, at times, let us blurt out a response that moments later has us groaning "How could I have said that?"

This fast limbic response ability reveals an important aspect of physiology. The limbic system, and hence all vertebrates, including reptiles, birds, and fish, store emotional information as long-term memory. And against this stored information, incoming data are analyzed. Memory is not the province of mammals alone. All vertebrates house within their brains some aspects of their personal histories.

Having raised trout in my high school years from egg through release into New England streams and lakes, I have no question as to memory in fish brains, notwithstanding the minuteness of their cerebrum relative to the size of their brain stem and limbic system. Just prior to each feeding of the fry, which were housed in outdoor pens measuring some ten to twenty square meters, I would whistle. By the second week, at the sound of my whistle, the water next to where I stood adjacent to the pen would boil as the fry thrashed about in expectation of the feed. Months after being released into lakes, the trout still returned at the sound of my whistle, though all feedings had ceased upon stocking. Similar techniques are used in raising salmon in fjords, and in training homing pigeons

Though long-term memory is not the possession of mammals alone, we still must ask if the response at the limbic level is at all conscious, or are the limbic contributions totally instinctual in animals other than mammals, and possibly in mammals also. Is consciousness solely provided by the cortex, and therefore a domain occupied by mammals alone? Did the fish "realize" that they were expecting food? Did they picture the food? Even without words, did they ponder what might be in store for them at that feeding, ruminating among a range of emotions tied to the flavors of the different feeds they had received in the past? Or did they experience only a direct instinctual response, just as our salivary glands react in a visceral way when we smell pot roast or hot chocolate chip cookies?

As the visual impulses leave the thalamus for the visual cortex, specific aspects such as color, motion, shape have been individually grouped. Now we encounter a lesson in an intricately woven division of labor, and the puzzling reconstitution of that division into a seamless, unified mental image, all within one-thirtieth of a second.

The visual cortex has six distinct layers, each devoted to analysis of an aspect of the image. These were identified over a century ago through the work of two pioneers in neurophysiology. Camillo Golgi, in the late 1800s at the University of Pavia, developed techniques of selectively staining nerve tissue. (Golgi apparatus bear his name.) For the first time it became possible to trace the path of a given neuron. His method is still in use. Santiago Ramón y Cajal, realizing the potential of Golgi's breakthrough, studied the nervous systems, especially the visual pathways, of many animals. In 1906, they shared the Nobel Prize in physiology and medicine.

Neural recycling time is about 30 milliseconds. Changes more rapid than this blur together. That is why a movie or video appears as smooth action. The visual cortex cannot distinguish between individual frames projected at a rate that exceeds the recycle rate. But within each of these slots of a thirty-thousandth of a second, each cortical layer analyzes a specific aspect of the scene as carried to it by the optic ganglion of nerves. In one layer, three-dimensional depth relations are studied by comparing differences from each eye, in another layer colors, in another motion; position; line orientation; and boundaries and contrast between boundaries. The nerve cells in the layer that become active in response to motion show no interest in the color of the object in view. In general an object moving into view elicits a stronger neural response than one moving out. Conversely, those neurons in the layer responsive to color, that is, the specific wavelengths of light, are not "interested" in aspects of motion. The individual layers feed their analyses back to the primary visual region and then via some unknown mechanism, at some elusive site, a unified image emerges.

If the entire visual cortex is destroyed, then no conscious sense of sight remains. The person is blind. However, if the eyes have not been damaged, signals may still reach the thalamus, the way station before proceeding to the cortex. The fast track to the amygdala remains operational, and emotional response to the "view" can arise which the viewer is in no way able to explain since no view has been "seen." It's clear that there's a lot more going on in your head than just what meets the eye. It is not surprising that so many of the texts on evolution eschew any semblance of a mathematical analysis of the theories that random reactions produced this ordered, information-rich complexity. When Lawrence Mettler and Thomas Gregg decided to add a few chapters on the mathematics of evolution in their book *Population Genetics and Evolution*, they brought Henry Schaffer on board. The math Schaffer brings to this totally secular text states clearly that evolution via random mutations has a very weak chance of producing significant changes in morphology. Of course this is exactly why you will search long and hard to find rigorous studies of probability in the works of Dawkins, Gould, or any of the other spokespersons for random evolution. Their approach to evolution is atavistic, a throwback to the time of Darwin, when cellular biology was assumed to be a rather simple affair of slime within a membrane. As we've seen, molecular biology has revealed that it is a mountain more than that.

And as if the complexity were not enough to break through the shell of materialism, there is history as well: the puzzle within the fossil record.

For three billion years, between the oldest fossils of life (bacteria and algae) at some 3.5 to 3.8 billion years ago and the first evidence of animals in the fossil record, 530 million years ago, the fossil record reveals a flow of life that remained one celled or at most groups of cells clustered into structureless communities. No appendages, no evidence of mouth or limbs or eyes.

And then with no hint in the underlying (older) fossils, an explosion of complex animal fossils appears bearing the basic anatomical structures of all phyla extant today. It is what is termed by the scientific community the Cambrian explosion of animal life. Among those structures are eyes. The earliest eyes arrived with stereoscopic positioning, and with lenses that by their fossilized shape appear optically perfect for seeing in water, the habitat of those early animals. We just struggled through the complexity of vision, from the conversion of incoming radiation inducing electrochemical pulses to the analysis of those pulses of information by the host animal. How did all this complexity develop in the blink of an eye?

Especially confounding is the current similarity of the genes that regulate the initiation of eye formation among all five phyla that have visual systems. Were there in the fossil record any hint of a common ancestor of these five phyla that showed a nascent eye, the similarity would be explained as having arisen in that earlier animal. But there is no animal, let alone an animal with a primitive eye, prior to these eye-bearing fossils. Random reactions could never have reproduced this complex physiological gene twice over, let alone five times independently. Somehow it was preprogrammed. This inexplicable complexity arises over and over again.

The deepest part of the brain, the brain stem, is the first stop for impulses rising from the spinal cord. (See Figure 9.) This, in conjunction with the cerebellum, an organ tucked beneath the cerebral mass at the back of the brain, keeps our vital life support systems on track. Here pulses regulate breathing, heartbeat, and in part contraction of the smooth muscles that line the gastrointestinal tract and all blood vessels other than capillaries. With these two organs, we subconsciously tune our muscles. Something somewhere must regulate the force by which we close our jaws, or place our foot on the ground as we walk, or lift an egg. These organs perform those jobs, all with no conscious

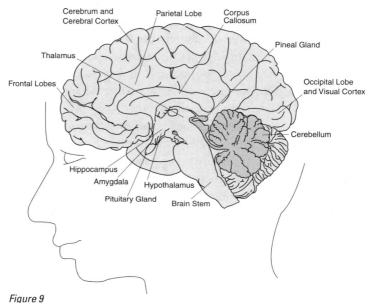

Figure 9
The Human Brain and Its Principal Organs

sensation of control and yet all precisely controlled. That is why when you take a step and suddenly discover that the ground or floor isn't where you cerebellum expected it to be, you get a jolt and lose your balance. Your foot hits bottom a lot harder than it normally would. The cerebellum had set the motion for what it "thought" was to be a standard step only to find out in the jolt that the situation was far from standard.

Your loss of balance resulted from an upset in the inner ear, another terrific device of ingenious design. Here we find an array of fine hairs set in three fluid-filled semicircular canals, each of which faces in a different direction, set at right angles (90 degrees) to the other two. This arrangement "feels" motion in any of the three directions. Tilting or rotating motions of the head cause the hairs in these canals to bend as the hairs' inertia delays their motion a minute fraction of a second relative to the motion of the head and canal itself. The hair's deflection induces a chemical reaction in a nerve at the base of each hair that initiates the electrochemical signal carrying the information to the limbic system. There the brain subconsciously integrates these data with information related to body position such as angles of leg joints, visual signals, muscle contractions, puts it all together, and learns that its body is falling.

Of course, the simple phrase "the brain integrates these data" hides a junglelike web of elaborately orchestrated neural activity with all the wonder of each nerve's mechanism for signal transmission, a billion and more axon/dendrite terminals in play.

For most of my life I've felt a transcendence within nature, some spiritual rumbling. My family had an apple orchard, and in my youth my father and I would tend it on weekends. The yearly arboreal cycle of winter hiatus giving way to the blossoming growth of spring and summer instilled in me that sense of wonder. But I had not realized the imminence of the marvel, its very presence within my body. Knowledge of molecular biology brought the story home.

Nature is so clever. To intensify the effect of motion, each individual hair in the ear canal is topped with a small node of calcium carbonate that adds weight and hence adds inertia. On a vastly different scale, it is inertia that pushes your head, like a huge node atop your neck, back onto the headrest of your seat as you zoom away from a traffic light.

Interestingly, this arrangement of hairs is insensitive to motions that are uniform, that is, motions without noticeable acceleration. When you wait at a traffic light and then suddenly notice that you are rolling forward, you press on the brakes, only to realize that it was the car next to you rolling backward that gave you the false impression of forward motion. In this case the visual sensation of motion overrode your inner ear's sense that all is stationary. There's more than one voice in each of our heads.

The brain stem is considered to be the oldest part of our brain, oldest in that it is found even in the simplest of animals in our phylum, the chordata. Yet it functions crucially in vision, a most advanced organ, as it activates the twelve muscles, six to each eye, that direct the eyes' movements in precisely synchronized fashion. Of course, recalling that the oldest of animal fossils house fully developed binocular vision, there is no surprise at finding ocular control in this ancient but not so primitive portion of the brain.

The entire brain, including the cerebellum, is composed of two hemispheres, right and left, with all organs but one being duplicated in structure, though not necessarily in exact function. As with nerves from the eyes, by the time we leave the brain stem, crossover has occurred. Stimuli arising from nerves on the left side of the body are recorded and processed in the right hemisphere; stimuli from the right travel to the left. Why this is so is a question yet to be answered, though it may be to match the inversion in vision forced by the shape of our eyes' lenses.

The one organ in the brain not present in a pair is the pea-sized pineal gland, located atop the thalamus. Being unique, it was once thought to be the seat of the soul. Most recent research implies it may be one of the seats of wakefulness. Its product is melatonin, the so-called sleep hormone. Bright light sends a signal to the pineal gland, causing it to decrease its production of melatonin, and in doing so, resets our body's clock to the wake-up mode.

The stimulus to speak, to state what is on your mind, is housed in the right hemisphere. Following a stroke on that side, the person may have much to say but getting it out from the brain into words will now take concentrated and seemingly conscious effort by the injured party. It also takes considerable patience on the part of the listener.

My mother, at ninety-one, could still construct and tell a good joke—slowly—though her left arm and leg were highly impaired by a right hemisphere stroke two years earlier. Formulating successful humor requires relating the logical to the illogical in a surprising way. That skill remained largely unimpaired. It's a lesson in relationships. A person confined to a wheelchair may still be a very full person inside her or his head.

Hearing, the mechanism by which sound waves are converted to electrochemical nerve impulses, is not one bit less complex than the transition of light into neural signals. Sound waves, which are composed of vibrating molecules of air, impact the tympanic membrane, the eardrum, which moves a chain of three tiny bones that move another membrane that sends the vibrations into a fluid-filled canal lined with tiny hairs. These hairs move in response to the changing pressures of the transduced sound wave. Their motion induces an electrochemical signal that races off toward the thalamus. (Notice the similarity between this and the sensing of balance, which takes place in an adjacent portion of the ear.)

Perhaps the most exceptional aspect of the hearing process is the rapidity with which the twenty thousand hairs of the ear can respond. Middle C vibrates at 256 cycles per second. The C above middle C is 512 cps. The C above that is 1,024 cps. The hairs must be able to resolve these fantastically high frequencies in order to differentiate among notes in each of these ranges. This is by far one of the fastest rates of response in our bodies.

In addition to resolving sound wave frequencies, the ear and brain must parse a stream of sounds into words and the words into bits of sentences, notwithstanding that each speaker has her or his own pace of speech, accent, pitch. For the most part we do it seamlessly, with not a conscious thought about the amazing interpreter we have within our heads.

Let's follow the path when, for example, a child's cry is heard at night. The long thalamus route sends the sound signal to the temporal cortex, located on the sides of the brain. The sound is deciphered step by step. Is it a noun or a verb? Is it part of a string of words, or does it stand alone as an exclamation? Then, what sound is it?—A voice. A child's voice. You have a long-term memory of that type of sound. It is familiar. One of your kids. The chain is now complete. The temporal cortex now knows that your child is calling for help and sends this information to the amygdala, which has already received a hint of the emergency via the direct route to the amygdala's subconscious, emotional memory. In response, it has already induced preliminary reactions, such as adrenalin influx to get the body energized for moving. From the amygdala, the call for action heads to another limbic location, the hypothalamus and the cerebellum, which goad your sleepy body into organized motion. You stumble off toward where you think your child's bed is located. Your memory, which is speculated to be a pattern of previously established axon-dendrite synapses somewhere within the cerebrum, having been initially laid down through actions of the hippocampus—also a part of the limbic system—provides this information.

Isn't anything simple in biology? The answer is no. Our every act is comprised of miraculous biochemistry.

And you thought you were just responding to your kid's call. You were, but not in quite the direct fashion your brain told you about it—or more accurately stated, not quite how you consciously perceived it. The brain doesn't bother your conscious mind with all the minute analyses involved in deciphering the signals, analyses that might make cryptology seem simple. All that is performed by the other you, the you you never ever meet. But it is there, housed inside the same head that lets "you" hear your daughter and see a rose and smell its fragrance. All held by what seems to be nothing other than a hundred thousand million axons, each having thousands of terminals, connecting and interconnecting with a million billion (1,000,000,000,000,000) dendrites. To get an inkling of what that number, and hence your brain, is all about, I urge you to count to a billion, a million times. After all, in some sense, the part of your brain you have not met has done it, so why not have the you you know do it also? At one number each second, with no breaks for resting, that task will occupy you for the next thirty million years. (Considering that amount of time, we'd better send out for coffee and doughnuts.)

Of course, a nineteenth-century view of the world can justify the archaic belief that it all evolved by random reactions among atoms. That belief was conceived before molecular biology opened the Pandora's box of hidden complexity.

The brain has space for two versions of you: the you you never meet but that meets with you every moment of your life as it regulates all the automatic functions of your body; and the you you know so well, the one that feels as if it is just above the bridge of your nose within your forehead. The you you know is also a composite of two: the analog emotions whose source we often cannot even identify, and the particulate sensory data of sight, sound, touch, taste, and smell.

Some Eastern religions refer to that spot on the forehead as the third eye. It might be equally termed one of the three "I's": the logical I, the emotional I, and the I I never seem to meet.

Is it possible that, in parallel with this bizarre subconsciously lived multiplicity of our mind, there is also a multiplicity within the world, a world unrealized at the conscious level, but still very real in its impact upon the world our conscious physical senses can access? This would be metaphysical, in the sense of being outside the physical. Call it, perhaps, an underlying wisdom from which the physical world emerges. A physicist might call it information. If the parallel is complete, then both the physical and the metaphysical are together embedded in a higher singular existence. A theologian might call that singular existence God. A physicist might call it the metaphysical potential field that collapsed and gave rise to our universe. Our brain might be the sole organ by which we are able to sense the metaphysical.

There's anecdotal evidence that the physical is actually embedded within the metaphysical. We all have an unexplainable nebulous desire to reach for some higher purpose, for meaning, in life even after we've satisfied the survival needs of food, clothing, and shelter. The indefinable nature of that sensation is somewhat akin to that which my student encountered when he could not quite pin down why the scent of a particular aftershave aroused in him an avalanche of emotions he thought were long gone.

8

THE PICTURE

IN OUR MIND

*A specific combination of selected atoms allows me to type a letter,
say the letter* d, *on my computer keyboard and have the letter* d *appear
on my computer screen. With some systems, I can also choose the color
of the letter. If I'd majored in computer science or electrical engineering, I'd understand the circuitry that set the path from keyboard to illuminated letter on the screen.*

*A specific combination of selected atoms, mostly different from those
of the computer, allows me to think of a letter, say the letter* d, *and the
letter* d *appears as a mental image. If I so choose, I can envision the letter in color. But there is a fundamental difference between the* d *on the
computer screen and the* d *pictured in my mind. I can have studied
neurophysiology for years on end and I still would not have a clue as to
how that image gets mentally displayed.*

*The modular construction of the brain has allowed identification of
the exact cortical regions from which specific mental images arise. Unfortunately, knowing from where they originate tells us nothing of
how they come together to form the pictures we see in our minds. It reminds me of the case with electricity and a host of other phenomena. I
learned long ago that protons and electrons have equal and opposite
electrical charges. Unfortunately neither I nor anyone else has a clue as
to* why *the phenomenon of electrical charges exists. Yet these forces
form the basis for the organization of all material existence. We take
them as givens, as inherent characteristics of nature, something like
the images in our minds.*

We've reached the last stop in our investigation of the brain: how we physically perceive consciousness.

For most of the miracle we refer to as life, the brain is an on-going story, one of continuous neurological growth and attrition, or "pruning" in the jargon of the profession. Within the first few months following fertilization, genetic coding establishes the basic structure of the brain. That fixes the locations of its organs. The genetic hegemony over brain structure is clear. All humans have their vision processed at the back of the brain, language on the side, logic in the front. But the absolute genetic control stops at the level of the modules' locations. Once those are established, a lose-win arrangement begins in earnest. A sort of mental Malthusianism, if such a word exists, takes over. Each of the modules produces a surfeit of neurons and axons, sending them out to the general target areas, but not hardwiring them into specific locations in those targets. Of course, just to reach the general area is a wonder yet to be understood considering the distances these axons must traverse relative to their micron-sized dimensions.

Once in place, stimuli generated spontaneously by the organs themselves and also received from the environs—visual, auditory, tactile—cause these axons to fire action potentials. As this neural activity forms patterns, some connections are reinforced, others are ignored. Those axons that consistently fire in unison, and therefore are likely to be receiving similar stimuli from a common source, wire together. This enhances their common synaptic connections. Those axons that are not being actively stimulated are "pruned." Use it or lose it is an adage that applies to the brain as well as to muscle tone. We may have close to six feet of body below our heads, but it is through the sum product of those connections in the brain that we know our bodies and ourselves.

Let's take a quick overview of the cerebral terrain, and then return to crunch through the details. It's the wisdom implanted within the biology of the brain that lays the basis for the wisdom of the mind.

The brain stem, a three-inch-long organ, joins the spinal cord with the rest of the brain. A slight shortening of the stem has been related to the occurrence of autism. The lowest part of the stem, the medulla, regulates such thoughtless processes as blood pressure and breathing. Just imagine what life would be like if every breath of air were a conscious effort. It's interesting that during speech we subconsciously override the brain stem's autonomous impulse to take a breath. This allows us to control air flow over the vocal cords. Perhaps it is because controlling our breathing touches such a deep part of the brain that conscious deep inhaling and exhaling is such a relaxing exercise. It is a very rare person who can consciously control the other domain of the medulla, blood pressure, but it has been known to happen.

Next up on the brain stem, the pons handles much of the change in body functions between sleep and wakefulness. Next in line comes the midbrain portion of the stem. Here the muscles of the eye are activated. Because this function is so very deeply seated, the eyes of a nearly brain-dead person can follow someone as she or he walks across the room.

Once above the brain stem, we reach the limbic system. We've already dealt in detail with its components and their functions. Briefly, the thalamus receives all incoming stimuli other than smell and routes them to the nearby amygdala for rapid emotional response and also to the cortex for slower logical consideration. As a crucial relay station, the thalamus connects with all major cortical regions: vision, sound, and motor (motion), helping to integrate the emotional with the rational. Adjacent to the thalamus, the hypothalamus, a cluster of pea-sized nuclei, connects with the frontal and temporal lobes as well as with the thalamus and the brain stem. Here hormones are secreted and, among other processes, feelings of hunger and satiation are induced.

The cerebellum, located at the lower rear of the brain, has its own cortex, though it is only three layered, whereas the cerebral cortex, the seat of our logic and conscious intellect, has six layers. The cerebellum helps regulate balance and the inherent sense each of us has of where the various parts of our bodies (limbs, head) are located relative to one another at each moment. Do you just take it for granted that you know where your feet are right now? Don't. It takes a brain to keep track of what is doing what and where it is doing it. If not, we'd be stumbling over our own feet. The brain stem gets most of the cerebellum's output.

Over the limbic system and cerebellum come the two hemispheres of the cerebrum. Outermost, covering the entire cerebrum, is the gray matter, the cerebral cortex, an eighth-inch-thick landscape of rolling hills (known as gyri) and valleys (sulci). Though the simple surface area of the brain is about a twentieth of a square meter, the folding brings its effective surface area to 1.5 square meters, a thirtyfold increase in brain power with no increase in head size. A brilliant solution that human logic might have overlooked. The cortex is the brainy part of the brain, the location that adds up the multitude of neural inputs, and, depending upon their summed magnitude, determines if they warrant further thought or action. Moving into the cerebrum, below the cortex, the gray matter gives way to white, colored by the myelin sheathing of neural axons that fill the entire region. Here, within the individual axons, nerve pulses are shuttled at breakneck speed in all directions, linking the various cortical lobes and joining the emotion of the limbic system with the logic of the cortex.

Deeply buried nuggets of gray matter, the basal nuclei, lie adjacent to the limbic system. Their axons extend to the nearby thalamus as well as out to the cerebral cortex. The basal nuclei help coordinate physical movement, monitoring neural initiation and termination of muscle contraction. When they are unable to fulfill their tasks, simple tasks of life become emotionally painful burdens. Malfunction induces the erratic, uncontrolled tremors known as Parkinson's disease.

From the deeply buried region of the basal nuclei and limbic thalamus, a veritable fountain of white-sheathed axons wells outward, radiating to and from all regions of the cortex. This is the internet of the brain. Much of the interior mass of the cerebrum, which itself makes up most of the brain's volume, is composed of these axons. Information transfer is a major job of the brain.

Neurons in the cerebral cortex make our decisions related to judgment and voluntary movement. They are our intelligence, all housed within those few surface millimeters of brain tissue. They decipher sounds into language, radiation into vision, store particulate, factual information—the cologne was by Armani—and via axons reaching up from the thalamus, they integrate this factual information with the amygdala's emotional memory of the smell. "The scent makes me happy but I can't say why" is your amygdala speaking to you via your cortex. Every thought, though not every emotion, emanates from the cerebral cortex. Each process has its own particular region within the powerhouse of cell bodies and dendrite synapses we refer to as the cerebral cortex.

The brain is a wonder machine of intelligence based on biochemical reactions, each expressing wisdom that in no way is presaged by the components from which it is built. At every level of life, from the isolated cell to the interaction of nerve and muscle, through to the 10^{15} neural connections within a brain, a depth of information surfaces that annoyingly has not an iota of justification being there. Nature, left to itself, favors disintegration, homogeneity. But the saga of life is a puzzling story of increasing complexity, of uniqueness, of order being locked in place, defying nature's degrading pull. And the brain is the top-of-the-line example of this successful struggle against oblivion.

The frontal lobes, encompassing the forward half of the cerebral cortex, account for the intellectual you. It's what tries to make you think as a human. Associative reasoning, the ability to form analogies (which is possibly the key ingredient of genius), conscious thought, speech, control of impulses—all find their origins here.

Then at just about the midpoint of the head, at the aft portion of the frontal lobes, the precentral gyrus controls voluntary movement. Integration of what we see with how we reach for what we see takes place here. So much of what we assume as natural, such as picking up a cup of water and moving it to our lips, emanates from this region of the brain. Every act we do, trivial though it may seem, must be controlled, and for the most part the cortices of the cerebrum and the cerebellum orchestrate that control. A deep cortex-lined valley, the central sulcus, marks the end of the frontal lobes and the beginning of the parietal lobes. Here touch sensations received from the body are recorded on one of the several mental body maps, each of which is inverted, heels over head. From here also arises the ability to recognize and manipulate symbols such as numbers for math and words for language. Along the lower sides of the cerebrum, at about the level of the temples and ears, the temporal lobes interpret the words heard in speech. The left temporal lobe deals with word recognition; the right with the ability to say the words—speech. It was the stroke's damage to my mother's right hemisphere that made it difficult for her to tell the jokes that she held within her left hemisphere.

Because of this division of labor between the hemispheres, there is a crucial need for transfer of information from one side to the other. The cross-linking of information between the hemispheres is accomplished by a massive bridge of axons, some eighty million, the corpus callosum plus additional axon bridges among the individual lobes within each hemisphere. Were it not for the intense cross linkage, the two selves within each of us would be battling for control at the conscious and subconscious levels. But transfer alone is not sufficient. To avoid potential right-side/left-side conflicts, the information must be shared in a timely fashion. This requires synchronization to within some sixty thousandths of a second. Considering the complex, multi-faceted signal processing intrinsic to vision and hearing, this precision in timing is nothing less than astounding. For vision, if this were not the case, the reconstruction of what one side of our eye "sees" would be out of phase with that of the other side. The brain would find an overlapping or doubling of the view. For speech, we'd be stumbling over our own words, hearing echoes generated within our heads. Once again we come face to face with the brilliance of nature.

Seeing, reading, hearing, thinking words each has its own dedicated section in our heads. When speaking, which involves motor control (mouth and tongue manipulation, vocal cord tension, breath control), neural regions in the precentral gyrus and parietal lobes fire. Seeing words in a written text engages the occipital (visual) cortex. Thinking and mentally forming words activates the frontal lobes. Clearly in the design of the human brain, language carried a lot of weight (see Figure 10).

Major cortical areas used for motor control (movement) and sensation (especially touch) by chimpanzees and bonobos, the animals genetically most similar to humans, in humans are de-voted to language, especially that part of language related to formation and implementation of speech. Language is, in essence, the encoding of information. The intellectual processes of language in humans have replaced the physical ro-bustness of movement found in other primates.

It's by no coincidence that the oldest extant interpretation of the Hebrew Bible makes the quality of communication the dis-tinguishing characteristic between humans and all other ani-mals. This relates to work in the second century, when the Bible was translated from Hebrew into Aramaic by the scholar Onke-

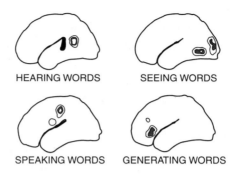

HEARING WORDS   SEEING WORDS

SPEAKING WORDS   GENERATING WORDS

Figure 10

Responses of the Human Brain to Different Tasks Related to Words

The most active areas, according to brain scans (PET), are shown as dark areas in this figure. As you read these words, this is what's happening in your brain.

(Figure after Marcus Raichle, School of Medicine, Washington University)

los. In Genesis 2:7, we are told that the soul of humanity, the *ne-shama* in Hebrew, was instilled in humankind. In modern terminology, we'd refer to this event as the change from homonid to human. The verse concludes in its usual translation, "And the Adam became a living soul." The Hebrew however contains a subtle difference, reading, "And the Adam became to a living soul." Based on the superfluous "to," which in Hebrew signifies a transition, Onkelos wrote, "And the Adam became a speaking spirit." Not merely speech marked the transition from the pre-Adam, humanlike hominids discussed in the Talmud to the spiritually full human Adam. Anyone who has raised animals realizes that animals communicate. The change with humans was the special spirituality that entered the communication, a spirituality that could only be communicated by speech as we know it.

The essential role of the frontal lobes in character determination was demonstrated through a tragic accident to a nineteenth-century railway workers. Phineas Gage and his colleagues were laying tracks, using explosives to help clear the roadway. Explosives in the wrong hands can yield unexpected results. As they set a charge in place, premature ignition sent a steel rod smashing through the front of Gage's face. It entered just below his left cheekbone and continued upward through the top front of his skull. Miraculously Gage survived—physically. Mentally, he was a changed person. His prefrontal cortex, the area just behind his forehead, the part that in essence is the logical self, was mangled. And the self Phineas now confronted was no longer the self he and his colleagues had known before the tragedy. Formerly a responsible, even industrious employee, he became a compulsive drinker, most often incapable of completing tasks that required attention for more than a few moments. Yet his self-awareness remained, indicating either that the rod had missed some specific patch of the frontal cortex devoted to self or—more probable considering the extensive damage—that the projection of self is like that of a hologram. Destroying part of the lobe still leaves enough information to create the whole, though in a highly degraded form. Emotions became the major driving force in his behavior.

Today, with sophisticated brain imaging, we realize that Gage had lost a part of the frontal cortex having intensive linkage with the unconscious impulsive self of the limbic system. Since the latter remained unscathed (physically, though certainly not emotionally), the emotions of the amygdala could now at last run the self with considerably less moderating interference from the cortex's logic. In a word, Phineas Gage became somewhat of an impulsive child housed in the body of an adult male. Childlike because the axons that join the cortex with the limbic system complete their myelin sheathing, and hence their information-carrying ability, only during the late teenage years. Until then the impulses of the limbic have a major say in behavior. Gage had lost the target nerves in the cortex of those connective axons.

Children are usually treated more leniently by the law than are adults who commit similar crimes. There is a neurological basis for their social behavior. But what of Gage? Is he responsible for impulsively breaking the law? His will had been so severely compromised by the physical damage that, while it might be advisable to isolate him from open society, it would not seem proper to punish him. We have learned from Gage and others with brain lesions in the frontal area that much, perhaps all, of one's personality is composed of neurons and the multitude of axon/dendrite synapses they form. The private, individual self we each hold so dearly as our very being is sequestered in quite discrete portions of the physical brain. It is not some mystical metaphysical cloud that surrounds our head.

The arrangement of those neural patterns through which our conscious mind reads the self is built gradually from each life experience. These form our memories. Simply said, in the words of psychologist Hanna Shir, "You are your memories." From studies of brain-damaged persons, a map elucidating the physiology of memory is slowly emerging.

Memories and the emotions they bring to the cortex provide the information upon which we base the choices of our free will. According to the Bible, there is a third input to the choices we make. That is the soul of humanity, the *neshama*. The *neshama* looks at the choices presented by the logic of the brain and analyzes each in relation to whether the result of the choice will move us closer to or further away from the universal consciousness, the unity that pervades all existence. It's the *neshama* that urges each of us to seek meaning and purpose in life even after we've satisfied our natural drives for survival and pleasure. The *neshama* is the creation mentioned on day six of Genesis that changed an animal, amoral Adam into an Adam with moral responsibility (Genesis 1:26, 27). Phineas Gage lost none of his *neshama* in the accident, but he had lost his ability to formulate within his window of free will those choices based on logic. The *neshama* can act only on the choices with which it is presented. Phineas remained a human, though he had to be treated as a human child housed in the body of an adult.

CP routinely drove her young daughter to school. Monday, the 13th of October 1986, was no different. By coincidence the date coincided with the 10th of the Hebrew month of Tishri, the holiday of Yom Kippur. A car swerved, a collision occurred, a seat belt failed, and CP's head slammed into the windscreen, bounced off, and smashed onto the side window. Nine-year-old Leah was unhurt. Following the accident, CP remembered nothing of her life that preceded the crash. To her, Leah was just another child. She had to relearn speech. Interestingly, most of her motor skills remained unimpaired. How to walk, the use of cutlery, were all still there. These types of voluntary movements are coordinated unconsciously, like the skill to ride a bike even after not having been on one for years. They are mediated in the cerebellum. This "little brain," located behind and below the rear of the cerebrum, has its own cortex, and hence its own in-house capability for dealing with routine actions. Memory of love was a different matter.

The actual cause of CP's impairment was not clear from brain scans. What we do learn from experiences such as hers is that aspects of memory are distributed over the brain, and that while access to parts of long-term memory may be lost (though possibly still present in neural connections no longer able to be reached), new memories can still be deposited and then stored as long-term. CP was able to remake a full life, though as a person quite different from that which Leah and her former friends had known. Not everyone is so fortunate. A classic case is that of Barry Tiller (not his actual name).

Barry had suffered for over a decade from serious debilitating seizures caused by erratic neural activity in the cerebral area in and about the temporal lobes. At the age of twenty-seven, Tiller in desperation opted for experimental surgery. (The operation was performed during the early 1950s.) The medial portions of both temporal lobes and the forward (anterior) two thirds of the hippocampus were removed. His seizures ceased. But so did his ability to store all new memories for more than an hour. Unlike CP, who lost all record of her former life but could build new memories and a new personality, Barry Tiller retained access to the old but could add nothing new. He was totally unable to transfer short-term memory to long-term. Each day's experiences were brand new. Doctors who visited him daily were always new acquaintances. A novel was out of the question since it would take longer to read than Barry could mentally store the plot and characters. An hour after lunch there was no memory of what had been eaten.

Every new experience we have, whether it's choosing tonight's menu or looking at a work of art, is embedded in a seamless flow of consciousness that relates the past to the present. In our thoughts and fantasies, awareness of self and the passage of time are always present. For Barry Tiller, the passage of time ended in the 1950s. He could not fathom that forty years later he had become an old man. After all, in his mind he was still twenty-seven.

But Tiller retained the ability to store one type of newly acquired information. That was the ability to learn new motor skills, skills that did not require conscious or cognitive awareness in their execution, like riding a bike. He could even learn to play tunes by rote on the piano. Once newly learned, they were his thoughts to keep. Just as CP had not lost her subconscious voluntary motor skills, Barry Tiller could learn newly acquired skills of this type.

From these and other similar (though less dramatic) cases, two distinct regions for laying down long-term memory have been identified. For those memories requiring strong cognitive, cerebral cortex involvement, the hippocampus is essential. The memory does not reside in the hippocampus, but the hippocampus is the way station that actively takes the short-term memory and transfers it to long-term storage. With the hippocampus gone, as in the case of Tiller, the brain has no way of transferring short into long term. The exact location of the storage bin or bins is still a matter of conjecture, though it is thought to be somewhere in the cortex. The route for implanting cognitive memory must not be the route for its retrieval. Tiller could retrieve with full capacity events that occurred prior to his operation, even though he no longer had most of his hippocampus. For motor skills that are memorized and then performed later by rote, the cerebellum with its own cortex appears to be up to the task of storing and handling their implementation.

Between 90 and 95 percent of all persons are right-handed, right-eyed, and right-footed. The hemisphere skewing toward the left does not appear to be socially inflicted. By the fourth month following fertilization, most fetuses already suck their right thumbs, a left-hemisphere activity.

The right brain is not without its skills. It knows its way around. Lost in a city or a forest? Listen to the right lobes for spatial clues as to where you are and how to get out. Severely damage the right side of your hippocampus and you will get lost inside your own home. Facial recognition is also a subconscious facility of the right hemisphere. It is in fact quite similar to using spatial clues for navigation. But, as with navigation, though the ability is built-in, the implementation for a given region or face requires learning.

For several years beginning in 1981, I traveled frequently to the People's Republic of China as an adviser to their government. In those early years just after China was reopened to the West, my flight from Israel was quite circuitous, the requirement of the diplomatic reality of the time. (It was common knowledge that I had Israeli citizenship. China had not yet recognized Israel politically, though it admired Israel's technology.) I had no trouble envisioning the plane's roundabout flight plan, but disembarking was always for me a moment of extreme embarrassment. I found it a near impossibility to recognize the persons whom I had met time and time again. My mental plan for facial recognition was, and remains, so firmly fixed into the Caucasian mode that I never learned to distinguish clearly among the dozen or so smiling Asians who waited to whisk me through customs. They had no trouble spotting me on the tarmac. In those days I was the only Caucasian in the crowd. To my Chinese colleagues I was the different face in the crowd.

The right brain has more extensive long axon structures than the left brain. It integrates the whole, while the left brain likes details, calculations, abstraction of mathematics. The right brain seems to be the more emotional side, picking up rhythm as opposed to the digital notes and words. As with language, music is also processed in parts: rhythm, pitch, loudness. The reconstructed piece yields an emotional as well as an informative response. It's Beethoven (fact) and it's beautiful (emotion); it's Satchmo and it swings. The complexity of the processes lends credence to the notion that our brains are wired for music.

In the same temporal region in which the right brain hears music, the left brain is devoted to language comprehension. The location, known as Wernicke's area, is named after its discoverer, the nineteenth-century German neurologist and psychiatrist Carl Wernicke. In the same period, the French surgeon and anthropologist Paul Pierre Broca identified the location of deliberate speech generation. The area, now named after Broca, lies in a posterior portion of the left frontal lobe.

Logically, language comprehension is processed in a region adjacent to the left hemisphere auditory cortex. This shortens the physical connection between the cortical region where incoming sounds are recorded and the place where those sounds are deciphered into language.

The right brain seems to tap emotions related to pessimism, depression, and fear, while the left brain actually has a region—in the mid upper frontal lobe—that when electrically stimulated evokes a feeling of amusement to the point of laughter, unhumorous as the current situation might be to the unstimulated brain. The temporal lobe may be the site of spirituality. At the least it is the site of perceiving spirituality. Stimulation there can yield what has been described by patients as a divine experience. Neurologically speaking, the reaction is similar to temporal lobe epilepsy. Vincent van Gogh is thought to have experienced the affliction intermittently.

Is a joke truly funny when we laugh at it, or is it merely some aberration of our frontal lobe? My guess is the surprise juxtaposition of the logical with the illogical is actually that which our brain experiences as funny, notwithstanding the ability to get a laugh electrically out of a totally deadpan situation. We see sparks of light when we bump our heads into the door in a pitch-black room, and also when the visual cortex is electrically stimulated. But we also see light when photons strike our eyes' retinas. Like the joke we find funny, light really exists, notwithstanding our ability to induce the sensation of light by a bump to the head or a shock to the cortex.

The same is likely true for feeling the divine spark. That we are physically wired in our brains to experience spirituality, that a spiritual sensation can be induced by electrical stimulation to a part of the brain, makes an inspirational moment or event no less real spiritually. The biblical message of Eden is that humankind was created for pleasure. That we blew it is another matter. If pleasure is part of the message for life, being wired to appreciate a good joke fits right in. If apprehending the spiritual in the commonplace is also part of the message for life, then it is equally logical that we would be wired to apprehend the divine.

. . .

143

THE effects of testosterone on the fetal development of the brain become evident in adult behavior. Though both women and men have the same general brain structure, and also left-hemisphere dominance, processing by the parallel lobes differs significantly. Neural connections between the two hemispheres have been reported to be larger in women than in men, both at the corpus callosum and at the connections between the two lobes of the cerebellum. The junctures facilitate the integration of the emotion and rhythm of the right hemisphere with the logic and language of the left hemisphere.

In standardized tests of skills, women in general do better at matching items having common characteristics. Men excel at rotating three-dimensional objects mentally. The language skills of women exceed those of men. Women develop language at a younger age. Men generally prove better at abstract math-like reasoning, while women are better at the details, at arithmetic, and at highly precise manual tasks. Men are better at archery, at hitting a target, catching a ball. Men can learn a travel route faster. Women are better at remembering landmarks along the way.

From twenty years of experience as a teacher and supervisor of prekindergarten play groups in villages and kibbutzim, Yael LaHav has found that from the first day of attendance, the girls head for the dolls and playhouse, the boys for the jungle gym. Of course by the time they are three or four years old, society may have imprinted this behavioral pattern. It stays for life. Just picture a hunting trip. Do you envision men or women? And whom do you mentally see at home, tending the kids, keeping house, men or women? The answers are obvious. In the few hunter-gatherer societies still remaining, the male-female split of tasks is the same. Molecular biology and brain imaging show that much of the mental difference between the sexes is nature, the imprint of hormonal activity on the prenatal brain, and not nurture, the pressures and morays of society. If IQ is a measure of overall abilities, then women and men are similar.

Considering the spread of individual talents between the sexes, if each of us is willing to accept that being different does not mean being less valid, less valuable, less intelligent, then it seems a woman-plus-man team makes a very good combination for survival in a world that moment by moment poses a multiplicity of problems each begging its specific solution. That seems to be what nature had in mind with sex and also what the Bible had planned from the start. "And the Eternal God said, 'It is not good for the Adam to be alone; I will make a helper opposite him. . . . Therefore a man shall leave his father and his mother and cleave to his wife and they shall be as one flesh" (Genesis 2:18, 24).

9

THINKING ABOUT

THINKING

TAPPING INTO THE CONSCIOUS

MIND OF THE UNIVERSE

Within the brain we perceive the consciousness of the mind, and via the mind we can touch a consciousness that pervades the universe. At those treasured moments our individual self dissolves into an eternal unity within which our universe is embedded. That is the message both of physics and of metaphysics.

Is a single molecule of water, one unit of H_2O, wet? Not sure? Well, what about two molecules? Are they as wet as water? What's the point of transition at which a batch of H_2O molecules become wet? What about a single neural axon/dendrite synapse? Does it contain a mind?

As for the wetness of H_2O, from its chemical properties, the hydrogen bond, and the 104-degree angle at which the two hydrogen atoms orient as each shares an electron with the single oxygen atom, we could predict that if there were enough H_2O molecules present, we would find the properties of what we refer to as wetness. Wetness arises from the interactions, the weak bonding, among billions of billions of H_2O molecules. Each drop of water falling from a faucet contains 10^{21}— that's a thousand billion billion—molecules of H_2O.

And what about the axon/dendrite synapse? Is it a bit of a mind?

Research has proven that a synapse can be the site at which a specific piece of information is stored. If we knew all the properties of this single synapse, we might be able to predict the vast analytical potential of joining similar synapses by the billions into an interactive, associative network. We could foretell that the primary capabilities of such a union would be the storing and processing of information. We might even call that union a brain. But unlike the prediction of the wetness of water, we'd have not an inkling that this brain, at some level of complexity, say at a concentration of a hundred billion neurons, each exuding a thousand or more axon terminals, would go critical, like a nuclear reaction, and give rise to the mind we see within our brain.

The mind is our link to the unity that pervades all existence. Though we need our brain to access our mind, neither a single synapse nor the entire brain contains a hint of the mind. And yet the consciousness of the mind is what makes us aware that we are humans; that I am I and you are you. The most constant aspect of our lives is that we are aware of being ourselves. Even in the illogical jumble of a dream, filled as it may be with fantasy, the constant is that we are ourselves.

The brain is amazing. The mind even more so.

A few years ago, IBM's supercomputer Deep Blue was taught the ways and bi-ways of chess. The AI folks at IBM had built a very clever machine. In the spring of 1997, Deep Blue challenged chess master Garry Kasparov to a match. It was a battle for both of "them." For a while it looked as if Kasparov might win. But Deep Blue carried the day. In time, and likely in short time, an invincible program will beat all human challengers, always, and probably with relative ease.

That victory neatly laid to rest a claim made by Berkeley philosopher and professor Hubert Dreyfus. In his book *What Computers Still Can't Do*, Dreyfus predicted that AI—computer-generated data manipulations—would never reach a level of skill able to outwit and outplay a chess champion. It took less than five years for Silicon Valley to prove Dreyfus wrong. Kasparov's defeat carried a strong message to Dreyfus and others of his ilk. To paraphrase the late Harry S. Truman, if you don't know how to cook, stay out of the kitchen. If you're not skilled in computer science, don't make predictions about what AI can and cannot do. If you want to philosophize about the capabilities of AI and the wonders of the human brain, then first learn the technology and the biology. It is for exactly that reason, learning the terrain, that I took you through the workings of the brain. I want us to address the brain/mind interface from a stance of knowledge, not from some deep-seated commitment to an emotionally charged heritage, be it materialist or metaphysical.

For all the well-deserved hoopla over Deep Blue's impressive victory, a major accomplishment for AI, for all its phenomenal data-crunching, algorithmic power, I doubt that Deep Blue was aware of even a pinch of the emotion that Mr. Kasparov experienced as he sweated through the tourney. And much of the brain's function is exactly this type of mathematical robotics. We can make a machine that laughs just as we can find within the brain the cortex module that when stimulated forces a guffaw. The mechanical operations of the brain can be matched and outmatched by AI. What makes the accomplishment of Deep Blue brainlike rather than mindlike is not the skill of a chess move. It's the lack of the experiences that Gary felt as a human battling a machine. The brain takes facts and integrates those facts with emotion. The mind takes the product of that integration and experiences it. A computer will probably never reproduce what Heidi felt as she watched the sun say goodnight to the mountain snow. That's the work of a mind.

SCIENTISTS may someday prove that the mind is totally a flesh-and-blood, physical phenomenon, or, quite possibly, they will find it emergent, not defined by the physical. From what we have seen of brain function, whether the mind is purely physical or partly metaphysical—call it divine if you wish—the brain's very existence is quite simply mind-boggling and quite possibly miraculous. There is brilliant design in the brain, and to make it requires the nature of our universe, which means we need a metaphysical force, a potential not composed of time, space, or matter that created the time, space, and matter of our universe. It's worth reemphasizing: the inequality between cosmology and theology is not whether there was a metaphysical creation. That is a given. The debate is whether the metaphysical whatever-it-is, or was, that produced our universe manifests interest in the physical reality it created.

The entire processes of the big bang and the evolution of our universe are more fantastic than any science fiction buff ever imagined. It sounds almost biblical. Not just the elusive physics of the process, but the fact that the effects of this single event are felt till today, fifteen billion years later. The entire universe, we included, formed from, and runs on, the energy brought forth in that ancient flash of creation. That galaxies and people and orange juice and oatmeal could come from a ball of energy should tell each of us something about the metaphysical force that brought it and us into being.

When our kids were younger, they would ask Barbara and me about our lives in the United States, before we moved to Israel. They wanted so much to rummage through the steamer chests of my parents and grandparents, piled in the attic of what is now a two-hundred-year-old farmhouse on rural Long Island. We, as adults, are not so very different. As we now contemplate the universe's birth and growth, we are rummaging through what are the remnants of our cosmic childhood. And at each stage we meet head-on a reality that is so brilliantly designed and inexplicably complex that we simply have to take it as a given.

Why should matter attract matter? Why doesn't matter repel matter? Why should protons and electrons have their equal and opposite charges, even though their masses differ by a factor of over one thousand eight hundred, and a neutron have no charge even though it has a mass quite similar to that of a proton? Why the Pauli exclusion principle and the quantized characteristics of orbital electrons? All these realities are humanly illogical and totally arbitrary, but without them there'd be no molecules, no rocks and water, no brain, no mind.

What does it take to make a brain? It takes a big bang producing a universe guided by laws of nature somehow tuned to lead energy into rocks and water on a user-friendly planet that can take those rocks and water and change them into a marvelously complex, data-crunching, algorithmic, sound-, sight-, touch-, and smell-sensitive wonder capable of processing thousands of inputs in parallel with a cycling time of thirty thousandths of a second. When contemplating the amazing complexity of the human life form, don't just think of the entire body encased in a smooth flexible layer of skin. The skin hides the wonder within the body, just as visible nature masks the metaphysical within which it is embedded.

Think of the inner workings. Think triple-layered cell membranes with voltage and protein-regulated channels for getting nutrients in and products out. Think RNA polymerase receiving a signal from some remote region and then searching for and opening just the correct spot on the three-billion-nucleotide-long helix of DNA, pulling complementary nucleotides from solution to produce mRNA that, when transported out of the nucleus, will find a ribosome that will decode it, pulling molecules of tRNA bearing just the right amino acid out of the twenty variations of amino acids in the cytoplasm, all at fifty shots a second. Each cell producing two thousand proteins every second. Think motor molecules carrying their protein cargoes step over step along microtubules. Golgi apparatus, neurotransmitters, and more and more. Multiply that complexity by a billion written out a hundred times for the brain alone, and then for each of those hundred billion nerves, sketch out a thousand axon/dendrite synapses.

The wind blows and thousands of leaves shimmer in the sun. Your eye sees them all. A million, more probably a billion, ion channels opening and closing along the ganglia of a million optic nerves leading from retina to thalamus and on to the visual cortex, cycling thirty times a second, as bioelectric signals, the information that records the motion of each of those leaves, reach into your brain. A myriad of chemical reactions, all in parallel, simultaneously recording the data.

Trees, eyes, the brain, from inert rocks and water via dumb unthinking random reactions? Logic alone tells you it could not have happened by chance. But the materialist superstition of our culture, the idea that if you can't measure it, it's not there, insists that chance be the explanation. And once a fact is imprinted on a mind, like the song a sparrow learns in its youth, that fact is yours for life. Believe it or not!

And yet here we are. A small part of a vast universe thinking about its origins, rummaging through steamer trunks in the attics of space and brain, trying to find the meaning of that which we call the mind.

FOLLOWING his defeat by Deep Blue, Garry Kasparov is quoted as having felt he was being challenged by a "new kind of intelligence." Indeed he was. The intelligence that confronted Gary was all calculation and no emotion. It was the cerebral cortex and more, but with no amygdala getting in the way. For Deep Blue, the match was truly no sweat. It's the perfect approach for chess, but for love it's a loser.

It's no secret that we can isolate within the brain those areas responsible for given activities. We can even measure the gradual increase in size of specific cerebral lobes we associate with cleverness in living vertebrates as we move from representatives of the oldest of vertebrates in the fossil record, aquatic life (some 530 million years ago), to reptiles (first fossil appearance 320 million years ago), then opossum-sized mammals (250 million years), primates (60 million years ago), and finally Homo sapiens (sixty thousand years before the present).

But the mind is very much greater than a layering of holistic feelings of self and awareness onto the observable facts recorded by the brain. True, the conscious mind arises from the brain. Destroy the cortex and you destroy consciousness. Destroy the brain and the palpable mind goes with it into oblivion. But the physical organs of the brain may be only the circuitry that makes the mind humanly perceptible. In that case, a form of consciousness may remain. Smash a radio and there's no more music to be heard. But the radio waves are still out there. We just don't have the apparatus to change the electromagnetic radiation into mechanical sound waves. The brain does for the mind what the radio does for music.

Brain-mind questions have all the trappings of disputes in which illogical solutions are required, yet resisted, to explain data that contradict established, out-of-date theories. The persistence of theories for a randomly driven evolution of life in the face of data from molecular biology and the fossil record, both replete with evidence against it, is one such example. Another, now largely settled, was the yearning in 1960s and 1970s for an eternal universe, one that would obviate the need for a metaphysical beginning.

Once a paradigm is established, only two avenues exist to depose it. Either a more logical paradigm is formed that fits the data satisfactorily, or overwhelming evidence demonstrates that the existing model is wrong even though no new model is known. The erroneous concept of an eternal universe fell under the weight of data that consistently indicated a big bang start of time, space, and matter. Just what the nature of the metaphysical force that produced the creation might be is still being debated. The tree of Darwinian evolution has bent and bifurcated to become a bush of evolution. That too fails to satisfy the known data, but there is no other *physical* model even remotely possible as an explanation and so randomness persists as the catechism of the school faithful to neo-Darwinian belief.

Modeling the brain/mind interface, forming a paradigm for how the brain gives rise to the mind, suffers a similar challenge. There is no hint of how we physically view and hear and smell the messages of the brain. Yet a metaphysical solution is untenable to a materialist school steeped to believe only that which can be seen or measured, the summing of the individually observable parts. Unfortunately, at the brain/mind interface, this reductionist approach misses the crucially holistic nature of the mind.

Quantum mechanics required a paradigm shift from classical mechanics, a shift even more extreme than accepting a universe with a beginning. Quantum mechanics necessitated replacing logical observable processes with the "illogical" phenomena of the subatomic world. The very existence of our universe calls out for a metaphysical explanation, an explanation that by definition is illogical in physical terms. The undisputed yet enigmatic existence of our self-awareness, our consciousness, does the same. The mind may be our only link to the reality of the metaphysical.

J. A. Wheeler likens the universe, all existence, to the expression of an idea, to the manifestation of information. This has the ring of quantum mechanics. The widely discussed quantum wave function that is attributed to each and every entity in the universe contains the information that totally describes that entity. The implication of Professor Wheeler's statement is not only that the wave function, which in street language refers to information, is a fundamental property of existence. The implication is much more profound: that information is the actual basis from which all energy is formed and all matter constructed. It sounds, at first cut, bizarre. But then, before Einstein's $E = mc^2$ who would have guessed that the basis of all matter, solid, liquid, gas, in every corner of the universe, is something as ethereal as energy, the totally intangible, completely massless, wave/particle duality we label as a photon? The massless, zero-weight photon gives rise to the massive weight of the universe.

That bit of strange, illogical physics is a proven fact. An equally illogical aspect of our amazing universe may be on the horizon. As specious as it may seem at this moment, if Wheeler and others are correct, information may be the fundamental substrate of our universe, a substrate made visible when expressed as the energy and material and space of the universe. In a strong sense, our universe may be the manifestation of information.

That swings back to a thought presented much earlier. Every particle is an expression of information, of wisdom. The self-awareness we experience is the emergent offspring of that wisdom. The more complex the entity, the more complex the information stored within. We tap into it via our brain. Because information is present in all existence, the consciousness I feel as my self-awareness has a cosmic history. It does not arise from my brain de novo. Aspects of it have been present from the start, the very start, the big bang. Consciousness, as wisdom, is as fundamental as existence itself.

The basic question is, just as ethereal energy can give rise to that which appears to be solid matter, can information provide the basis for that which precedes the energy? Such a suggestion is not so very outlandish. Rest assured, no one has a clue as to what makes a photon, the basic "particle" of energy.

The idea that a universal consciousness, emergent from this wisdom, might be present in what we habitually refer to as inert matter finds support in a range of scientific data. Entangled particles act in concert even though separated by distances that obviate the possibility of their communicating in the reaction time required. It is as if each particle is simultaneously aware of the other's action at the instant of the action. A classic example of this linkage is the famous double slit experiment in which waves or particles passing through an open slit, call it A, know whether slit B is open or closed. The particles behave as if they are conscious of the condition at B even though B in no way affects conditions at slit A.

In the Drakensberg in South Africa, my family and I watched from a distance as one baboon preened another, and then after a quarter hour or so at this activity rolled over and had the other baboon preen her (or him?). That certainly smacks of some sense of self-awareness. And it makes all the sense in the world, if consciousness at varying levels is ubiquitous.

A grain of sand contains the slightest hint of the skyscraper of which it is to become a part. Do the very elements of the brain, the carbon, hydrogen, oxygen, nitrogen, have within them the barest trace of consciousness, which will combine and emerge as the complexity of a fully functioning brain? The same elements that build our brains and bodies form stars and galaxies. In that case our minds are but a part of a vast and conscious universe.

The elusiveness of the mind finds a parallel in that of other fundamental givens in nature. The properties of gravitation, electric charge, and magnetism can be measured only via their interactions with or relative to other matter, never in isolation. Unlike a cup which when held remains a cup, we can't grab a photon and observe it as a photon—hold it in our hands and examine it from all sides. Every catch of a photon changes the photon into something else—heat, an energized particle. Whatever emerges from the interactions, the photon is gone, and like the smile of the Cheshire cat, only its interaction remains. Likewise, though we can examine and identify the various lobes of our brain, we can only sense our consciousness. We can't isolate it.

In classical physics, all future events are deducible from the initial conditions of the situation. Quantum mechanics revealed a contradiction to this age-old assumption, a contradiction highly illogical and equally highly relevant to thought. We can accurately predict the general flow of a reaction, but not the exact path. Because the flow of all future paths is what is called probabilistic, known in general but not exactly, the exact future is not contained within the present. This is one of the reasons for assuming time's arrow moves only forward and never reverses. The probabilistic nature of nature, the spread in possible outcomes, means we cannot reconstruct the exact past from the present. And so we cannot go back to the past. This is not chaos theory, a theory based on our inability to measure exactly the present conditions. The quantum reality is that there is no exact present to measure.

In our brains, this probabilistic spread relates to the range of thoughts from which we select our next action. In a sense, the divergence opens a window of options from which we may choose, either via the logic of our frontal lobes or the raw emotion of our amygdala.

Plants and humans are sensitive to sunlight. But plants always move toward the light. Their DNA programs produce this reaction. Humans usually prefer being warm and therefore enjoy the sunshine—usually. That is, though humans often prefer being warm, some choose to move toward and some away from the sun. Identical sensory input, sunlight, yields very different reactions in different persons. Current environment plus memory produce the decision. That's our frontal lobes at work, integrating the fact of heat, plus emotion, the feeling of being too hot or too cold. The decision is selected from among the possibilities stored within the synapses of the brain. The choice is ours.

Is being aware of the feeling that it's too hot or too cold different from the feeling itself? The puzzle is not that we react to the sensory inputs of a situation. The puzzle is how we conceptualize in mental terms what seems to be actual physical reality. When I think ice cream is cold, or chocolate chip cookies smell wonderful, I neither feel bodily cold nor smell the cookies, but I do recall the cold and imagine the smell.

Conscious self-awareness is what makes me know that I am a specific human. I know it is I and not you who am in my mind. At about two years old, children begin to realize self and that others have minds too, and that those minds can be manipulated. (Raising kids is so instructive as to how the mind works.) There are aspects of our selves that we can describe, such as professional skills, joys, fears. Looking at the data from a brain scan, a skilled technician may be able to locate the parts of the brain active during specific mental and motor tasks. A neurologist, knowing the brain terrain and seeing the data output, may be able to tell what the activity was that induced each specific brain lobe to fire. For one who knows the visual cortex very well, the data can even reveal if the scene being observed is in color or black and white. But the experience of the vision is something else. That escapes the machine's reading.

We talk about missing links in evolution. We have a missing link right in our heads at the brain/mind connection. The move from brain to mind is not one of quantity—a few more neurons and we'll tie the sensation to the awareness of it. It's a qualitative transition, a change in type. The mind is neither data crunching nor emotional response. Those are brain functions. Mind functions are self-experience, seeing, hearing, smelling. The replay of what came in. These are phenomena totally different from the acquisition of the information. That is why adding up the synaptic data would predict a brain, but not a mind.

There is an aspect of science known as Fourier analysis. When presented with a very complex mix of data, such as several individual waves superimposed one upon the other, using Fourier analysis we can strip away those waves that represent noise and be left with the pure information that we seek. Can we strip away the sensory and emotional data of the brain, and be left with the experience of those data, the mind?

Sir John Maddox, former editor-in-chief of the renowned journal *Nature*, summed up our knowledge of consciousness in a piece featured in the December 1999 issue of *Scientific American*: "Nobody understands how decisions are made or how imagination is set free. What consciousness consists of, or how it should be defined, is equally puzzling. Despite the marvelous successes of neuroscience in the past century, we seem as far from understanding cognitive processes as we were a century ago." It's not likely that we'll find a mind by looking in a brain, though we require a brain to contemplate the mind. Fourier analysis does not help in the realm of sensation.

Defining an aspect of our world in an arbitrary manner rather than in absolute terms is not new to science. Essentially all of scientific inquiry is based on fundamental "givens," values that just are what they are for no discernible reason, such as the electromagnetic charge. These facets of nature are observed as being intrinsic to the working of nature, but arbitrarily set by the laws of our universe. Gravity is not matter and protons are not charge. Yet gravity emerges from matter and charge from protons. The universe might have been created without gravity, without electric charge. It would be a very different universe, but a universe nonetheless. Some givens are totally illogical in human terms of logic.

The quantized nature of radiation is but one example in which a new and fundamental component, the quantum, had to be introduced (by Max Planck in 1900) to explain an observed aspect of reality that has no a priori explanation. That the flow of time should pass at different rates because of relative velocity or differences in local gravities is patently ridiculous. Time is time. The duration of my minute should take exactly as long as the duration of your minute no matter where you are. Makes all the sense in the world. But as Einstein predicted in his revolutionary theory of relativity, it just turns out not to be true. The rate of flow of time varies from place to place in the universe. Sometimes we just have to fold our reductionist tents and realize the whole may not be totally contained in the sum of the parts. Consciousness has all the trappings of another nonreducible element of our universe. The conscious mind is not mystical, but it may be metaphysical—meaning out of the physical.

The discovery of nonlocality, of action at a distance, illogical though this phenomenon is, has revealed the linkage of disparate parts of the universe. The infinitely extended wave characteristics of all matter give physical basis to the metaphysical claim that the entire universe is entangled. Taken together, these seemingly illogical but validated insights point to something quite logical and marvellously wonderful. The universe is truly a uni-verse. All existence is joined through the expression of information, an idea, wisdom. Our mind is the emergent link that occasionally taps into that unity. You know when it happens as the surge of exhilarating emotion envelopes your entire body. At those moments, as one's local individuality dissolves into the unity that embraces all existence, we realize the full meaning of "the Lord is one."

ILLUSIONS

GAMES THE BRAIN PLAYS

WITH THE MIND

My computer seems to have a mind of its own. When I press the return key, it insists upon indenting the next line of text even though this time I want lefthand justification. The brain outdoes the computer by a mile in getting its own way. Like a computer, the brain is a victim of habit. View two or three dashes and the brain images a line. The brain always wants to fill in the dots, even when dots are all there are. The difference between a computer and a brain is that you can eventually get your own way with a computer if you persevere. With the brain, there are situations when there is no solution short of surgery!

In essence the brain says to reality: Don't bother me with the facts. I have my preconceived notions. Ideas triumph over the physical world—even when they are mistaken.

We say seeing is believing, but what is red to your eyes may be maroon to mine. The difference in opinion is not merely one of definition. The difference is in how the brain images reality. The brain may actually "see" the colors differently, integrating its nature (DNA instructions for how the brain's neurons were distributed, largely before birth) and nurture (personal history; how those neural connections were altered and reinforced by experience). Together, nature and nurture conspire to provide consciousness with a finished product, the inside view of the outside world.

The breakthrough of the brain is that nurture actually can alter what nature gave us. What might take genetic mutations a multitude of generations over millennia to accomplish, nurture can bring about in seconds. The images and sounds and smells and touches that you let into your brain today actually determine how you think about and see the world tomorrow. What you think (or see or sense) determines how you think. In other words, don't expect to philosophize in the morning, if you orgy in the evening. As Louis Pasteur observed, chance favors the prepared mind.

But some mental prejudices seem oblivious to the nurture aspect of life. They are hardwired from the start. We are used to seeing illumination from above. That's the sun's usual position relative to us. For reading a book it would be hard to light the opaque pages from below. But what about face recognition? Lighting a face diagonally from below might do just as well as top lighting. But the brain's cortical regions for face recognition are strongly prejudiced toward shadows produced by light arriving from above, as with the sun. Lighting of even the friendliest of faces from below can produce a sinister, even grotesque expression, and sometimes a shape that is not even recognizable as a face.

The phenomenon known as lateral inhibition is equally hardwired in the brain. Look at the grid below of black squares on a white background. Notice the gray clouds that keep popping in and out at the intersections of the white lines? There is absolutely no gray present. None. But try as you may, you cannot stop your brain from "seeing" those gray clouds. To your brain, the pure white paper at those locations is gray. You recognize this as an "artificial" effect in the grid, but realize that at each letter you are reading on this page exactly the same phenomenon is taking place, making the letters appear more sharply in your mind's eye than they are printed on the page (see Figure 11).

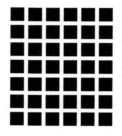

Figure 11

The effect arises from the eye/brain system wanting to present a mental picture in which contrasting borders are artificially accentuated. The brain is trying to help the mind see the boundary. As the visual signal travels from retina to thalamus and ultimately is dissected and recorded at the visual cortex, there is a progressive increase in border contrast at each stage.

The nervous system accomplishes this contour enhancement via lateral connections among adjacent neurons. As the action potential races along neuron A, which has "seen" light, it inhibits adjacent neuron B from firing. B not firing tells the brain that at the retinal start of B there was no light—hence the gray. Even if there happened to be some light at B, the brain will never find out about it. If an area is equally lighted, then all the neurons are equally excited and equally inhibited. The result for that situation is a mental picture of flat light. Only when there are inequalities does the inhibition become visible.

The subtlety of the process is amazing. Notice that the gray clouds appear only at the intersections of the white lines, and not along the white lines between the intersections. That's because the neural cells receiving stimuli from the intersection are surrounded by lighted cells on four sides, producing fourfold inhibition. Neural cells getting information from the white lines alone are inhibited from only two sides. Nothing other than surgery, which I am not suggesting, will remove the effect. It's built in from birth. Lobsters have the same problem!

But it's not really a problem. In most cases lateral inhibition helps us avoid bumping into things. The repetitive circular pattern below teases the brain in the same way, but here producing a sense of virtual motion between adjacent rings (see Figure 12).

Figure 12

Ernst Mach in the late 1800s produced virtual motion, a flapping effect, in his famous Mach figure (see Figure 13 on page 164). One moment the vertex points in and the next it pushes out. It could almost fly away, but it's all in your brain. Nothing is moving. Just a two-dimensional drawing that, because of the implied perspective and total lack of reference frame, makes the brain think it's three-dimensional.

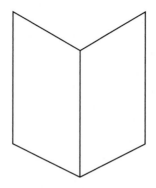

Figure 13

Then there is the question of is it a cup or two faces? A duck or a rabbit? Which of the lines is shorter, which line longer? (See Figure 14.)

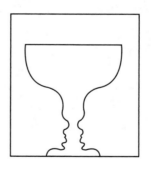

Figure 14 (continued at top of facing page)

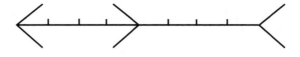

And don't tell me there is no triangle in the Kanizsa drawing below (Figure 15). It may be a phantom, but my brain tells me it's there. But what sees the triangle, the visual cortex or the visual mind? The former is physical, the latter virtual. The visual cortex records the data. The mind does the seeing.

So much for the naive thought that seeing is believing. The brain uses its past experiences, integrating them with current points of reference, to form its opinion. When those points are missing, the brain may fill them in. Data from brain scans indicate that it is the frontal lobes, the logic areas of the brain, that take the data from the visual cortex and decide what to forward to the conscious mind. In some cases the brain cannot make up

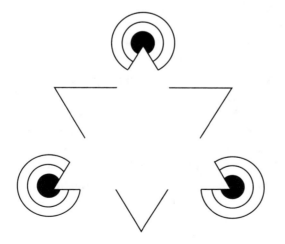

Figure 15

its "mind." Physicist Roger Penrose and artist M. C. Escher exploit these elusive neural effects to produce their famous drawings of figure/ground reversal, or a staircase that ascends and descends simultaneously.

Intellectually we live with figure-ground reversal continuously. The world looks so physical and yet I know with absolute certainty that its creation was the result of a metaphysical force. But what is physical? What is metaphysical? Where's the divide? Just what are protons made of? Don't tell me three quarks, because then I just have to ask what quarks are made of. It all reduces to fields of force, which means that for all the tightly packed hardness of a piece of rock or iron or any matter, there really is nothing there at all. Finally I can give a meaning to one of the more famous statements attributed to Yogi Berra: Nobody goes there anymore. It's just too crowded.

There are logical relationships for which the brain hunts, and when they are not there the brain does the best it can with the facts it has. Be they visual or verbal, the brain strives to have them fit within the context of what it perceives as logical. And logic is based on deductions drawn from the brain's personal history.

In front of me is a tree. As I change my gaze, rotating my head slowly to the left, the information that my retina receives of that tree moves toward the right peripheral visual field. That is the only information that the retinas get to send to the brain. So what should the message be? Good grief; the tree is moving. But the brain does not try to convince the mind that the tree is actually moving to the right, even though its image falls ever further to the right in my eyes' view of things. Why not? The cerebellum lets the cortex know that it's the head, and not the tree, that's on the move. In this case the brain corrects for the illusion that the world is moving about.

But, far from any view of the land, stand on the deck of a boat rocking so gently that the ocean swells are barely perceptible, and on a clear moonless night you'll watch the stars swaying back and forth. It's quite a shock at first, since as every landlubber knows, the stars are the "fixed stars." When you can't feel the motion, it's hard to tell what's actually doing the moving. I feel from every sense I have that it is the sun that circles the earth. Yet the earth on which we are standing is rotating at close to a thousand miles per hour, and we can't feel a bit of it. Inertial motion, motion without changes in velocity, eludes those specially weighted hairs near the inner ear that we studied.

Hold your hands steadily out to the side, fingers spread apart, pointing upward. Now try to count them while looking straight ahead. Tough going. You can't count them. Only a very narrow central retinal field, the fovea, has enough receptors to differentiate among items even as large as fingers. Objects off to the side become hazelike. But peripheral motion, that's a different story. Move your fingers one at a time and you can count them all, which means the information is there, but blurred together when not in motion.

Try reading a book while lying on your side, or watching a video while stretched out on the couch. Not easy unless you align your head with that which you are reading or watching. When a slide at a lecture is projected incorrectly, on its side, the entire audience as if by command tilts their heads to compensate.

Yet we can learn to read upside down. Jewish communities of European extraction read from the Torah with the scroll laid flat on a large table. The reader stands at the base. Kids in the congregation, mine included here in Israel, crowd at the opposite side of the table and read along, but to them the text is wrong side up. They read with equal facility from both views.

Perspective lets the brain construct the impression of 3-D images from the 2-D scenes viewed on a TV screen or gallery painting (see Figure 16). Nothing like a few points of reference to make depth plunge into a flat piece of paper. The brain does it all because it knows that as things get smaller, one behind the other, each is progressively further away. It's still a flat piece of paper no matter what the brain tells the mind.

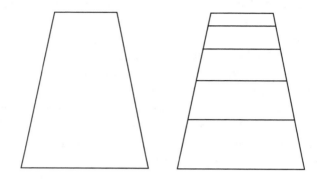

Figure 16

Perspective is what makes the sun and moon appear larger at the horizon than when overhead. The brain needs a clue to tell it whether these celestial bodies are nearby or far off. Only as they approach the horizon are such markers available. Being behind the terrain of the horizon means that they are beyond the horizon, and therefore they must be far away and therefore large. Swimming off the Israeli coast near Dor, at full moon I could see, with the turn of my head, both the moon just above the eastern horizon (moonrise) and the sun just before setting in the west. Both being near a horizon, they both looked large and of equal size. The sun's actual diameter is four hundred times greater than the moon's, but the sun is 93 million miles distant from the earth while the moon's distance is some 238,000 miles. This four-hundred-fold difference in distance accounts for the similar perceived size. The horizon presents no clue of these measures and therefore no telltale sign as to whether the sun or the moon is further or larger.

The eye has one obvious flaw. The nerves of the retina must

pass from the front back through the retina, in order to proceed to the brain. This punches a hole in the retina, a place where there are no light-receiving cones or rods, and in doing so produces a blind spot in the signal sent to the brain. But no such lacuna appears in your mental view of the world. That's because the brain fills in the space by assuming that what it sees around the spot continues into the blank. It's quite amazing to find this blank.

O   1

Hold the book about 9 inches from your face. Shut your right eye. While facing straight ahead, look diagonally at the "one" with your left eye and slowly move the book toward your face. The "O" will disappear. It does so when the image of the "O" falls on the blind spot of your left retina. As the book continues in, the "O" reappears as the image moves to an active site on the retina. You'd never know the blind spot was there without checking. The brain keeps this defect a secret.

Our brain gives us the illusion that we feel our toes in our toes. But every study makes it clear that all physical sensations, be they touch or smell or sound or sight, are registered in our heads, projected onto one of several body maps laid out on the cerebral cortex. So why do my toes feel like they are way down on my feet? Because that's the illusion my brain presents to my mind—a necessary and powerful example of the triumph of ideas over physical reality.

Daydreams may be wishful thinking, but they always make sense. Sleep dreams are altogether different. No logical processing occurs here, so any image we can conjure up slips by the brain's logic censor and appears on the screen of the mind. I wonder if when we are consciously aware, the mind may be driving the brain even as the brain presents its data and emotions. When we dream, day or night, the brain may be driving the mind. That we have several levels of consciousness is clear. How many times have you driven home, only to arrive and not remember much of the trip, but be mightily aware of the thoughts you were mulling over in your mind during the trip?

The ultimate illusion is one that nature plays on brain and mind. We look out at the world and see a myriad of solid objects. Scattering of high-energy particles by atoms tells us that the nominal radius of an atom's nucleus is 10^{-15} (one million billionth) meters. The nominal radius of the surrounding electron cloud is 10^{-10} (one ten-billionth) meters. That means the fraction of the total volume of the atom that is solid, a cubic function of the radius, is one part in a million billion. That number is a billion written out a million times. For every one speck of space in a solid iron bar that is filled with a nuclear particle, there are a million billion specks of space that are wondrously and gloriously empty. Like the blind spot in vision, there's nothing there no matter how solid it feels. Impossible to internalize, this fact is contrary to every experience we have. But it's a verified aspect of reality. Ethereal force fields among the atoms and molecules penetrate the voids, giving the vast open regions the impression of "solidity." If our eyes could perceive to the minuteness of the subatomic scale, we'd see we are walking on a very sparse grating. The illusion of solidness is due to the "weak" resolving power of our eyes.

Even the particles that make up the atom, the protons, neutrons, electrons, may not be solids after all. They may all be extended forms of energy. If indeed matter is the conscious expression of information, then the idea of mind over matter requires a revision. It must read the consciousness of the mind over the consciousness of matter. Anyone who has witnessed the holder of a black belt in karate shatter a brick while barely touching it (referred to as a soft break) will find nothing new in this idea. It's done more by concentrated thought (Chi) than by physical force.

Fighting it all the way, we are being dragged, kicking and screaming, into accepting the truth that our material existence is more fiction than fact. I say it. I teach it. The logic of my frontal cortical lobes analyzes the data and believes it. But in my limbic emotions, I fight it all the way. I want nature to be natural, natural by human, physical standards, and it doesn't seem to be turning out that way after all.

Even without cranial illusions, we find that the world gets curiouser and curiouser with each new discovery. Physics has touched the metaphysical realm within which our physical illusion of reality is embedded. In crossing the threshold from the physical to the metaphysical, science has discovered a reality it had previously relegated strictly to the mystical. It has discovered the presence of the spiritual, for that is really what the metaphysical is, within the land of the living.

AND what about the land of the nonliving? Though there is no explicit mention in the Torah of life after death, there are oblique references to what appears to be an aspect of life that extends beyond our physical existence. When the death and burial of a biblical personage are discussed in detail, the sequence is always the same: death, gathering to the deceased person's people, and then burial. Gathering to the people always occurs immediately after death and always precedes burial. In some cases, such as with Jacob, burial was months after death and gathering (Gen. 49, 50). Clearly, the gathering was related to death and not to the physical act of burial in a family sarcophagus. There's an intriguing aspect to this description of being gathered to one's people. This is exactly the experience that persons who have survived clinical death report. Deceased relatives reach out in vain to greet them as they return to the land of the living. Certainly this similarity may be mere coincidence. At near death, the brain must be encountering drastic chemical stress that might trigger these exotic "dreams." The concordance between the biblical account and the human experiences is however worthy of consideration, even though there is no way of proving which explanation is correct. (As an aside, I have interviewed several persons who had such experiences. Four claimed never to have heard of the effect prior to their encounter with it. When I asked one what she now thought about dying, she said she was not looking forward to it, since "it might not be as pleasant the next time.")

IF you take one thing away from this chapter, let it be that illusions are powerful. Sometimes the brain tricks the mind in its perception of reality, and sometimes the mind realizes the brain is up to tricks. Be that as it may, the brain and the mind working in concert have deciphered the most startling of all illusions: that the physical world we see about us is a structure so ethereal that were it not for forces produced by theoretical, never observed, virtual photons, we would fall right through, evaporate into nothingness, dissipate into a metaphysical cloud.

I think it was in *Duck Soup* that Groucho Marx declared: "Are you going to believe me or your lying eyes?!"

READING BETWEEN

THE LINES

WHAT DOES IT ALL MEAN?

There's an ancient tradition that when Divine revelation comes into the world, only one part is given as prophetic writings. The words are only a part of the message. The other part is placed within nature, the wisdoms inherent in the Creation. Only when we understand those hidden wisdoms will we be able to read between the prophetic lines and fully understand the message. With the help of science, we are learning to read between the lines.

We look at the world and see a marvelous creation. A myriad of forms fill the landscape. Diversity seems to be the message. But when we look below the surface, we discover a world made of a mix of identical particles that are actually waves and then realize that the waves are massless expressions of information. Physics has exposed the metaphysical basis of existence.

We study the physiology of a cell and the neurology of the brain, mapping the flow of each sound and each sensation as it makes its way from the world around us to its being recorded as synaptic unions among the multitude of nerves that store our intelligence. And then we envision that which we have seen or heard or smelled or touched. Activity increases in specific parts of the brain. A flow of images passes before our mind's eye. And therein lies the rub. We have not a clue as to how those images form in our heads, or from where they are played back. The mind is as ethereal as the seemingly solid floor upon which we stand.

We pick up a pencil, scratch an ear, and don't even dream of the billion and more complex biochemical/electrical reactions that secretly metamorphose our thought into neural activity and finally muscular contraction. We join with our spouse and nine months later, from information held within a cell and a half, the miracle of life repeats itself. But all the wonder is hidden. Why? Why aren't the subtleties of physics and the phenomenal symphony of which life is composed apparent for all to see, visible, right up front? Why are they sequestered beneath an exterior that looks so simple?

The implication is that we can settle for a superficial reading of nature if that's all we want, but the ultimate reality of existence lies below the surface, between the lines. The very fact that the Bible opens with the creation, followed by a detailed account of the physical development of the universe, carries a pivotal message. A single creation yielded all of nature, the energy, the matter, and the space for all future existence. It's not by chance that the only name for God used in the creation chapter (Gen. 1) is Elokiim, the biblical name for God as made manifest through the forces of nature. The message is that through nature, of which we are a part, we can discover the immanence of the metaphysical Creator within the creation.

That the essence of this metaphysical presence is not always obvious does not mean it is not actively there.

> And the Eternal passed before [Moses] and proclaimed: the Eternal the Eternal, God compassionate and gracious, slow to anger and abundant in kindness and truth. (Exod. 34:5)

That is a reassuring claim by the Bible, but is it true?

Purim is a holiday of costumes and feasting, of sending gifts and reading the Book of Esther. The holiday falls in that changeable time of year when winter and spring argue over whose turn it is to run the weather. Josh, aged five, dressed as a king, felt great as only a five-year-old can when he's sure his costume makes him look like the real thing. During the night before Purim, a rare snap of freezing rain had sheathed the trees of Jerusalem in ice. Frozen droplets hanging from the branches splashed the morning winter sunlight into a rainbow of colors. The air was like crystal, the sky blue.

We're not used to icy sidewalks in Jerusalem. The first I knew of the fall was when Josh's siblings came crying to tell me. Someone had left the door to our bomb shelter open. (Bomb shelters are a reality in parts of the world.) Josh had slipped on the ice and fallen down several of the rough cement steps. The disappointment of his torn costume matched the pain of the bump on his leg. As I hugged him, feeling his hurt and trying to make it all better, in my mind I raised a voice toward heaven and demanded in anguish: "Why did You let it happen?"

I wasn't expecting an answer and only later did I focus on the significance of my question. Not why did You make it happen, but why did You let it happen. Why didn't You step in and save the situation? Get involved, God. We just survived the twentieth century. Two world wars, the Holocaust with twelve million humans burned to death under the guise of a nationalistic ideology. Only to be outdone at the other end of the spectrum by the universalistic ideology of communism, with Stalin sending in excess of thirteen million persons to perish in the frigid wasteland of Siberia. From such horrors, it might seem that the putative Creator merely wound up the universe, gave it the energy to run, and then let whatever emerged play itself out. Such a scenario does not sound very much like the Sunday school version of the biblical God. If You are really there, God, what is Your role?

It is a question worth asking even by a believer. I wonder if

Abel equally pleaded with God as Cain beat him to death. In the famous "Am I my brother's keeper?" sequence, the Book of Genesis records God confronting murderer Cain: "Your brother's bloods cry to me from the ground." Bloods, in the plural. Apparently Cain didn't know how to kill Abel and so had to beat him repeatedly. There were many wounds. As Abel's life ebbed, blow upon blow, he must have called to God for help. But none came, though God had just accepted Abel's offering.

The Bible teaches reality, not some fantasy. Notwithstanding the biblical description of God as "compassionate and gracious, slow to anger and abundant in kindness and truth," the Bible knows that bad things happen to good people and God lets them happen. By the fourth chapter of Genesis, Abel, the good guy, is dead. Our impulse might be to flee from such a God. Jonah tried and failed. We are in this universe and that is the way it operates. If we were gods, our concept of how to run the world might be different. But we are not gods. (My guess is that if we were, considering the political and social problems we've generated, we'd make a much worse mess of it.) The biblical challenge is to understand the apparent randomness in nature, the acts that bring at times joy and at times tragedy, within the context of the claimed compassion and graciousness of a Creator involved in the creation.

Resolution of the apparent inconsistency between the immanence of the Creator and the presence of grief lies in the biblical definition of creation. In my previous book, *The Science of God*, I discussed at length the biblical description of creation. As learned from Isaiah (45:7), creation is the Divine act of *tzimtzum*, God's spiritual contraction. In the resulting spiritual space, the undifferentiated simplicity of God is fractured, yielding in its stead the variety of our universe. With this Divine contraction comes the release that allows for our choices of free will and the leeway for the seeming imperfections and meandering courses found in nature.

In the Book of Genesis, in a passage related to the patriarch of Judaism, Christianity, and Islam, we read, "And the Eternal said 'Shall I hide from Abraham that which I am doing. . . . For I have known him for the purpose that he may command his children. . . .'" (Gen. 18:17,19). Commenting on these verses, the thirteenth-century kabalist Nahmanides asked why is it written "I have known him"? Doesn't God know all persons? Nahmanides answered his own question. God knows all life, but the degree of Divine direction to an individual person depends on that person's individual choice of how close to God he or she wishes to be. A half century earlier, the medieval philosopher Maimonides made the identical observation in his *Guide of the Perplexed* (Part 3, chapter 51). Only the totally righteous have one-on-one Divine direction, and even that guidance may not ensure a life free of pain and suffering. For the rest of us, chance and accidents do occur. It's our choice as to where we, as individuals, fall within that spectrum of behavior that stretches from intimate Divine direction to total random chance.

Our difficulty in perceiving the metaphysical reality within which the universe is embedded is that we view creation from the inside looking out. We try to envision a metaphysical reality totally void of the distinctions of time, space, and matter. But all our thoughts are built of the physical images we have experienced, images of time and space and matter. Creation, however, is the exact opposite. It starts with a transcendental simplicity and gives rise to complex physicality.

Moses asked to see that transcendence but was denied permission. "And [Moses] said: show me please Your glory. . . . And the Eternal said you cannot see My face for no human can see Me and live. . . . You will see My back but My face you shall not see" (Exod. 33: 18, 20, 23). God's back is the imprint of the Divine within the world.

Einstein is widely quoted as having said, "The most incomprehensible thing about the universe is that it is so comprehensible." It is Einstein's discoveries that let us comprehend facets of the universe previously inconceivable. And it is those discoveries followed by the revelations of quantum physics that have clarified why the universe is in fact so very comprehensible to the human mind. It is because we are a part of the universe that has become aware. And in being part of it, we have come to comprehend ourselves. Our newly found self-understanding has revealed that wisdom is contained within even the simplest of particles, and an unfathomable and even perplexing amount of wisdom is present in the complexity of life.

The mystery of life's origins and its ordered complexity is not simply one more difficult scientific roadblock waiting for a physical explanation. Life, and certainly conscious life, is no more apparent in a slurry of rocks and water, or in the primordial ball of energy produced in the creation, than are the words of Shakespeare apparent in a jumble of letters shaken in a bag. The information stored in the genetic code common to all life, DNA, is not implied by the biological building blocks of DNA, neither in the nucleotide letters nor in the phospho-diester bonds along which those letters are strung. Nor is consciousness implied in the structure of the brain. All three imply a wisdom that precedes matter and energy.

Consistently, at every level of complexity, the information that emerges from a structure exceeds the information inherent in the components of that structure. This is true from subatomic electrons, the lightest of known particles, to the human brain, the most complex structure yet encountered in our universe. This ubiquitous emergence of information, of wisdom, cries out for explanation. The range and depth of it are fantastically unlikely to have happened by chance. Its origin is told in the three-thousand-year-old opening word of Genesis, *Beraesheet*. Not the superficial reading, "In the beginning," but the far deeper reality, *Be' raesheet,* "'With wisdom' God created the heavens and the earth." The substrate of all existence is wisdom.

As physicist Freeman Dyson stated when accepting the Templeton Award: "It appears that mind, as manifested by the capacity to make choices, is to some extent inherent in every atom. . . . God is what the mind becomes when it has passed beyond the scale of our comprehension."

And again, physicist R. B. Laughlin in accepting the Nobel Prize: "I myself have come to suspect that most of the important outstanding problems in physics are emergent in nature." Emergent, meaning not implied by the individual parts of the structure, not merely quantitatively different, but qualitatively; a difference in type from their components. The whole is not equal to the sum of its parts. The whole is greater than the parts could "imagine." And Professor Wheeler: the "bit" (the binary digit) of information that preceded and gave rise to the "it" of matter.

This emergence of wisdom in the material world appears as if it is de novo only if we fail to realize that an aspect of mind is "inherent in every atom." In a sense, mind has been present from the beginning. We might had known that all along. After all, with wisdom God created the heavens and the earth.

The link between the spiritual and the physical finds expression in the details of the biblical Tabernacle said to have accompanied the Israelites during their forty-year trek in the desert. For all the multiple listings of the items kept in the Tabernacle, the first item mentioned is always the ark within which lay the tablets Moses received on Sinai. The ark represented revelation, contact with the Creator. The second item is not the candelabra with its pure flame of spirituality, and not the altar with its message of sacrifice. The second item is always the table on which twelve loaves of bread were placed each Sabbath. The table represented the physical world, with bread as the penultimate symbol of the wisdom inherent in nature. To harvest a loaf of bread from the energy of the big bang requires far more than ideal laws of nature and an earth with a life-friendly environment. To get to a loaf of bread out of a big bang requires the emergence of the complex and orderly interactive system of information processing we find first in the simplest of biological cells and ultimately in the human brain and mind. Rather than explaining the appearance of this wisdom by attempting to establish the validity of the spiritual at the expense of the material, or the opposite, the material at the expense of the spiritual, the Bible in its wisdom by establishing a link between the ark and the table speaks of an integration of the two, the imprinting of the metaphysical upon the material. The physical and the metaphysical become one.

Humanity has spent and continues to spend billions of dollars on ventures such as the Hubble Space Telescope. We are seeking our origins with the passion that one lost in a desert searches for an oasis. "As a hart pants after water so my soul yearns for Elokiim. My soul thirsts for Elokiim" (Ps. 42:2, 3). Again, the Bible chooses the name of God as Elokiim, the aspect of the metaphysical Creator made manifest in nature. Our hope is that Hubble data will provide us with a clue as to the origin and place of life in the scheme of the universe.

Christian de Duve articulated the conundrum of life: "Faced with the enormous sum of lucky draws behind the success of the evolutionary game, one may legitimately wonder to what extent this success is actually written into the fabric of the universe." Being written into the fabric of the universe implies something other than merely having the created laws of nature amenable to the flourishing of life. Being written into the fabric of the universe tells us that we are made of the stuff of the big bang. We were present at the creation.

THE mind presents each of us with two inner lives. One deals with the demands and joys of daily life, and one seeks the transcendental, the eternal aspect of our finite existence. In all the thought experiments we might compose, fantastic though those imaginings might be, the one feature that remains is the flow of time. Without time all events cease, and without our memories, our concept of time vanishes. The world becomes a still life in the fullest meaning of the phrase. Time is the continuing reality of life. The relativity of time, discovered by Einstein, was the first of the steps that moved physics into the realm of metaphysics. I find it intriguing that time is also the first item that the Bible makes holy, holy in the biblical sense of being separate from the remainder of existence. Not a place, not a person, but totally abstract, intangible time—the seventh day, the Sabbath.

The Sabbath predates Moses, Abraham, Noah. Only Adam and Eve, the biblical parents of all humanity, predate the Sabbath. Long before the ritual of religion made its way into theology, the Sabbath was established. The Sabbath is the Bible's gift to all humanity; the crown of the six days of creation. It is the undersold superproduct of the Bible. It ritualizes contemplation, fits it into a timely rhythm, superimposing its cycle onto the other cycles that nature has imprinted through light and dark, satiation and hunger, phases of the moon.

The word Sabbath comes from the Hebrew *shabat*, meaning to rest, to cease from work. The essence of the Sabbath is rest. It stands in juxtaposition to the previous six days of work. Erich Fromm, in *The Forgotten Language*, described it perfectly: "Rest is a state of peace between man and nature. . . . Rest is an expression of dignity, peace and freedom."

The Bible understands the human psyche. It realizes that harmony between the two lives we live, the temporal wants of the body and the transcendent needs of the soul, is rarely a spontaneous happening. Without a ritualized, established routine there is always a reason for the tangible immediate demands of life to take precedence over our more abstract spiritual desires. There's no difficulty in being "holy" in a church or mosque or synagogue or temple. But the aspirations of theology far exceed our behavior in places of worship. The inherent aim is to bring the holy, the metaphysical, into the daily life of the marketplace. Bringing the spiritual into the tasks of the work week takes practice. Religion provides that practice. It's the pumping iron that gives us the spiritual strength to make theology a part of our mindset. The Sabbath is the day of practice. It's Eden, the message of which is that humankind was created for pleasure. The Sabbath returns to us a taste of Eden and helps us spread it through the entire week.

The absence of work introduces a formalized time for family intimacy on all levels. Not by chance, the structure of the Sabbath makes it the most physical and the most spiritual of days. A receptive spirit requires a happy body. Spirituality can't be weighed, but our emotions tell us there is an aspect to life that transcends the physical. Even after satisfying our physical needs, we humans feel there is still something missing. It's meaning. We seek purpose in life. The Sabbath opens space to consider what that purpose might be. The reality of quantum physics has proven that the future is not merely an extension of the present. There is a slack, a leeway in the direction taken by the forward flow of events. That margin of variability opens a window for how and what we choose, how we use our free will.

When we exercise our will, we choose from within a window of possibilities built from our cumulative knowledge and experiences. That window is not static. In a feedback loop, the range of our current choices is skewed by how we chose in the past. How we thought in part determines how we think. As we grow, our window of choices shifts in accord with our new experiences. Subconsciously, the brain presents a range of possibilities, some of which rise to the conscious. We choose from among them.

There are tools to help link physical needs to spiritual desires, our "is" with our "what ought to be," without denying either. If, for example, loving one's neighbor is a goal, then there must be some hints as to how to achieve that goal. There are, and they are listed in that same biblical passage of Leviticus (19:18) that insists we can learn to love: Don't take vengeance and don't bear a grudge. In daily terms, don't keep accounts of what your spouse did or did not do. If you want some poison for marital bliss, remember each time your partner was late or forgot your birthday or didn't compliment you on your new garment. That's keeping accounts. Acting on those accounts is vengeance. The Bible's suggestion is to define love as focusing on the virtues while acknowledging the shortcomings. Identify your loved one with those virtues. At the end of a tough day, consider that it may have been tough for your spouse too. Just before you walk in the door, review why you married in the first place. Your spouse is the same person, but now with the additional demands of family life attached, not all of which are filled with fun. Cut through and see beyond that baggage. A happy spouse is a happy house. Do it just for the totally selfish reward of having a happier life. Forget the spirituality altogether. That will come naturally, from the bottom up.

The human soul, the *neshama*, is our link to the transcendent. It knows the meaning of "the Lord is One." It realizes that there is a unity that pervades all existence and from which all existence is composed. The *neshama* looks at each potential act and quietly asks, will this act move me closer to or further from the joy of touching the oneness of creation?

As we learned from the tragedy of Phineas Gage, the brain can build upon and feed the mind only from that which it has stored. If we choose to feed it with trivia and violence, then rest assured trivia and violence is what will appear within our window. The tranquility of the biblical Sabbath prepares the mind for the spiritual, that part of our lives so often masked by the noise of the work week. One day a week, it says, expect pleasure. And the amazing truth of the human mind is that if you expect pleasure, you'll find it.

Fifteen billion years ago all of us and all we see were part of a compact homogeneous ball of energy. That was the entire universe. Once we were all neighbors. Then as the ball expanded outward, differentiation occurred, masking the underlying unity. Some minute fraction of the energy went into making the atoms of the ninety-two elements. I doubt if it could have been predicted which primordial atoms of carbon would end up in my brain and which in a leaf of the eucalyptus tree across our courtyard. There is plenty of room for chance and choice in the world. If I had been born a few hundred miles to the east, I'd be aiming missiles at where I am now.

Yet behind the chance, there's an absolute truth. It's the unity underlying all existence, the singular wisdom from which we are constructed. Your soul knows it. Listen carefully. You can hear it. With training you can learn to enjoy it. Sabbath R&R, rest and re-creation, is a good primer.

PLATO likened our perception of life to persons viewing shadows on a wall while unaware of the far grander reality that produced those shadows. Science has revealed part of that larger reality. In the wonders of nature, we have discovered the imprint of the metaphysical within the physical. As one who sees the wake of a boat that has passed by, so we encounter the hidden face of God. And the hidden face is indeed grand in its collective simplicity. We live in a time when finally we can begin to read the text that lies between the lines. That text is yielding to us the secrets at which the written words had only hinted.

Then the eyes of the blind shall be opened and the ears of the deaf unstopped. . . . The people that walked in darkness have seen a great light; they that dwelled in the land of images, upon them the light has shone. (Isaiah 35:5; 9:1)

EPILOGUE: HINTS OF AN EXOTIC UNIVERSE

Teacher of physics par excellence, the late Robley D. Evans, said that if you have key points you want to get across to an audience, first tell them what you are going to say, then say it, and then at the end tell them what you said.

We are at the end.

From the traditional view of classical physics, the universe seemed so logically constructed. Principles of determinism taught the obvious fact that identical causes produce identical results. There were the distinct and separate natures of particles and waves, and of energy and mass. All this fit well within our human psyche's opinion of how a world should work. And then came Einstein and Planck and de Broglie and Heisenberg, and the assumed logic underlying existence tumbled into an exotic, unpredictable fantasy that turned out to be true. Their discoveries revealed relativity, quantum physics, uncertainty, particles with fuzzy extended edges, particles that aren't ping-pong-ball-like entities but are actually waves. And then most bizarre of all, that these waves might actually be representations of something as intangible as information, as wisdom.

Enlightenment is far easier when we have no preconceived notions of what must be true. The social and professional pressure to conform to accepted ideas can be monumental even whenever mounting data contradict their validity. It's called cognitive dissonance. As great a mind as Einstein's bowed to convention when his cosmological equation of the universe showed that the universe might be in a state of violent expansion, that the galaxies might be flying out from some moment of creation. Rather than publish this astonishing prediction, he changed his data to match the popular, though erroneous idea that our universe was static. In doing so he missed the opportunity to predict the most important discovery science can ever make relative to the essence of our existence: Our universe had a beginning; there was a time before which there was neither time nor space nor matter.

Today we have another seemingly logical, but quite likely erroneous, piece of accepted wisdom forcing itself upon our paradigm of existence: that the physical world is a closed system; that every physical event has a correspondingly physical cause preceding it. It's not a question as to whether or not we can predict the exact effect of a given cause. Quantum physics says we cannot. But our logic insists that each physical effect must be initiated by a physical cause. How could it be otherwise?

The very knowledge of the big bang provides proof otherwise. The physical system we refer to as our universe is not closed to the nonphysical. It cannot be closed. Its total beginning required a nonphysical act. Call it the big bang. Call it creation. Let the creating force be a potential field if the idea of God is bothersome to you, but realize the fact that the nonphysical gave rise to the physical. Unless the vast amounts of scientific data and conclusions drawn by atheistic as well as devout scientists are in extreme error, our universe had a metaphysical beginning. The existence—if the word existence applies to that which precedes our universe—of the eternal metaphysical is a scientific reality. That single exotic fact changes the rules of the game. In fact, it establishes the rules of the game. The metaphysical has as least once interacted with the physical. Our universe is not a closed system.

But does that which created the universe still interact with the creation? That's a question each person must confront. Without solid scientific knowledge, philosophizing about this puzzle is very much a nonstarter. We've surveyed the science and discovered a complexly ordered wisdom expressed in the molecular functioning of life nowhere evident in the structures from which life is built or in the laws of nature that govern the interactions of those structures. That wisdom in life is the imprint of the metaphysical.

This may have the ring of a radically new idea. It's not. Thirty-four hundred years ago, the opening words of the Bible set forth the identical concept. *Be'raesheet*—with wisdom—God created the heavens and the earth. With a bit of intellectual endeavor, we can avoid the type of error that Einstein made as he bowed to a popular prejudice of his day.

Almost a millennium ago, the medieval philosopher Moses Maimonides wrote that we can know *that* God is, even though we cannot know *what* God is. "The Eternal knew [Moses] face to face" (Deut. 34:10). It is not written that Moses knew the Eternal face to face. "The Eternal said [to Moses and to us today]: I will make all My goodness [the wonders of creation] pass before you. . . . You shall see My back [as one finds, in the wake left by a boat, evidence of the boat's passage], but My face shall not be seen" (Exod. 33:20, 23). Even in the closest of encounters, the face of God remains hidden.

DNA / RNA : THE MAKING

OF A PROTEIN

Other than sex and blood cells, every cell in your body is making approximately two thousand proteins every second. A protein is a combination of three hundred to over a thousand amino acids. An adult human body is made of approximately seventy-five trillion cells. Every second of every minute of every day, your body and every body is organizing on the order of 150 thousand thousand thousand thousand thousand thousand amino acids into carefully constructed chains of proteins. Every second; every minute; every day. The fabric from which we and all life are built is being continually rewoven at a most astoundingly rapid rate.

The flow of life, like some cosmic ballet, is staggeringly dynamic. But the marvel life presents is held not merely in its vibrancy. Its wonder couches most deeply in the extent of the information that directs this vitality and in the fact that this information is qualitatively different from all other laws of nature.

To understand why your child resembles you, and why a stalk of wheat grown from a seed resembles the plant from which the seed was taken, is to understand how information is transferred biologically. That transfer rests in the exquisite functioning of two molecules, deoxyribonucleic acid—DNA—and its messenger/partner, ribonucleic acid—RNA—the superstars of all life. For three billion years, this molecular dream team has worked to produce every known life form, plant and animal alike.

Encoded on these two molecules as strings of specifically arranged nucleotides is the information dictating that humans have ears on opposing sides of the head, a face with a nose in the middle, a pancreas that produces insulin in response to the presence of glucose in the blood, and every other physical characteristic of our bodies. Humans have approximately thirty thousand such information packets, each packet known as a gene. Considering the intricacy of a human body, the number of genes is surprisingly small. And equally impressive is that all this is orchestrated by a set of instructions initially stored in a single fertilized egg.

The genetic revolution is still in its infancy and so we can only speculate on how the complexity of the human body is actually encoded. It would seem that for such a small genetic package to produce such meaningful structures, each gene must somehow provide a general plan as well as specific instructions. There may be codes within codes. The system by which genes are programmed in the DNA suggests this plausibility.

The DNA of each human cell contains approximately three and a half billion nucleotide bases. Clusters of these bases form the thirty thousand genes each of which in turn produces a specific molecule, most commonly a specific protein. Since some three thousand bases are used to code for a typical protein molecule, with over three billion bases available for the coding, this could produce more than a million proteins. Yet we make our living using approximately eighty thousand proteins. This means that the space occupied by the genes accounts for less than 10 percent of the total number of nucleotide bases. So why the other 90 percent? Is it just excess cellular baggage? Possibly—but equally likely is the possibility of genetic codes not yet discovered tucked within the remaining 90 percent that might instruct a cell how to build a human from the eighty thousand proteins. (That number is an approximation. The number is thought to be between 60,000 and 100,000, with a few estimates based on RNA sequences reaching 140,000.)

In this appendix I discuss the nuts and bolts of nature's greatest encrypted source of information, DNA, and how that information is held and read by the cell within which it lies. The process is nothing short of amazing. Along the way we'll discover some of the riddles that genetics has uncovered. Interesting though these puzzles may be, a much larger mystery looms below the surface: the source of the encoded information and the source of the awesomely complex, yet nearly error free, system that deciphers and translates this wisdom into the structures of life. In general, simple laws, such as the laws of nature, cannot give rise to complex information that exceeds their own unless that complexity is a fractal extension, a duplication in number and type of the base law. This is simply not the case with the genetic code. The information therein is apparent neither in the atoms and molecules from which DNA is formed nor in the laws of physics and chemistry that govern the interactions among these molecules. And yet if the fossil record is correct, the endowed wisdom of DNA seems to have been present from the very earliest stages of life on earth. How the coding that drives all life sprang into existence remains a mystery. The scale of the mystery is best realized by the complexity of its product.

Each human cell contains about two meters of DNA. Considering that a typical cell is nominally 30 millionths of a meter in diameter, squeezing in the two meters of DNA requires a feat of compression on a scale of one hundred thousand to one. Nature has the answer. Two complementary strands of the molecule twist and fold into a supercoiled helix, a helix within a helix, yielding a ball less than 5 millionths of a meter (approximately one one hundred thousandth of an inch) in diameter. To read the information stored, the helix is opened. But opening such a tight helix can cause damage. Therefore the bonding of the helix must be strong enough to maintain stability but sufficiently motile to allow rapid opening when the signal is given for information to be retrieved. A combination of strong covalent bonds maintaining the integrity of the basic molecular structure with weaker, more easily broken bonds, holding the strands in the helix, solves the problem.

As with the fine-tuning of the four forces of physics, the relation between the so-called strong and weak chemical bonds is equally precise. For basic stability, DNA is a winner. It has been recovered from hundred-thousand-year-old Neanderthal bones. The biological system of information storage is phenomenally dense. If all the information in all the libraries in all languages were transcribed into the language of DNA, it could be recorded within a volume equivalent to 1 percent of the head of a pin.

The entire encoding alphabet for the thirty thousand genes consists of four molecular letters, the nucleotides, laced along a repetitive molecular backbone. This greatly simplifies the production and selection of the materials needed to construct the DNA helix and its RNA messenger/partner. How these molecules work together to produce life is intimately related to their almost-too-good-to-be-true chemical structures.

Of this pair, DNA is the primary repository of the bio-information. To get a glimpse of this remarkable molecule of life, we must pass from the outer cell into the nucleus, a structure that appears as a cell floating within a larger outer cell. At some 10 millionths of a meter in diameter, the nucleus occupies slightly over 10 percent of the total cell volume. Its walls are perforated by thousands of porelike openings, each approximately 100 billionths of a meter in diameter. The micro-scale of it all is awesome considering that from this micro-world emerge fully formed humans and elephants and even great blue whales.

The nuclear pores are designed to keep DNA inside the nucleus while allowing ready access and egress for a host of other molecules. Interestingly, though the nucleus is built for production, an energy-intensive activity, it has no autonomous source of energy production. It relies totally on energy supply imported from the outer cell in the form of the familiar ATP.

And here's that enigma again. It's worth contemplating throughout the discussion because it shows its head in a dozen different ways, the problem of how the entire process originally got started. The first stage in making ATP requires a dozen or so intricately dovetailed protein enzymes, each one picking up the action just as the previous one leaves off. These enzymes are manufactured using information stored in the DNA. Retrieving and deciphering the wisdom held within the DNA in order to make the enzymes requires a good deal of energy. Get the problem? For the energy we need ATP. For ATP we need enzymes. But to make the enzymes we need the information held in the DNA and to get that information we need the energy supplied by the ATP. I guess if you buy a car at the top of a hill, you can drive away without a battery. It runs by itself. But something had to get the car to the top of the hill in the first place.

Of course the naysayers will call for another source of energy that started it off and evolved into ATP. But those pleas begin to ring thin. There is no hint of this evolution within the cell and usually nature keeps its ancestral record of development, such as the fetal heart "evolving" in the womb from an initial single tube to the mature four-chambered organ, or the human fetus at an early stage bearing a yolk sac (such as do fish larvae) that is then absorbed by the fetus.

And then there is the problem that, according to the fossil record, it all developed so very rapidly, almost simultaneously with the appearance of liquid water on earth.

I wish I knew what I meant when I agree with my colleague Dennis Turner that there's a ghost in the system. I'm a scientist. Studying nature is what has put bread on my family's table for a good number of decades. I want nature to work like nature. But at several key stages in the development of our universe, nature seems to have behaved most unnaturally. It's what Nobel Prize–winning physicist, and avowed atheist, Steven Weinberg referred to in his excellent book *The First Three Minutes* as the "embarrassing vagueness . . . the unwelcome necessity of fixing initial conditions," of having to accept a batch of initial conditions simply as "givens." "Givens" in scientific jargon is the sophisticated way of saying that that's the way it is and so let's start the discussion from those givens without understanding how they got there. First we start with the need to accept a universe housing forms of matter and the laws of nature as givens, for no a priori reason. And now at the other end of the dimensional scale we have DNA simply showing up on the scene with immediate complexity and working marvelously well.

The relationship between DNA and RNA is even more of a marvel. Through the pores of the nucleus, a continual supply of raw materials arrive bringing copious amounts of ATP and the building blocks needed to manufacture the information messenger, mRNA. In the opposite direction, passing from the nucleus into the outer cell, flows a steady stream of power-spent ATP, now referred to as ADP, and newly manufactured mRNA molecules, each carrying with it the design for, and the command to make, a specific protein.

Though DNA stores the information that makes the cell run, it functions in a purely passive mode, providing the patterns for mRNA molecules. The command to read those patterns comes from instructions brought to it by inducer proteins (again, proteins needed to make proteins, like a story without a beginning) arriving from the outer cell, and sometimes even from outside the cell. To get a feel for the almost unreal complexity of the cell's mechanics, let's first look at the structure of a DNA molecule, and then survey how it works. The basics are the same for all plant and animal life.

DNA comes in sections called chromosomes. Humans have forty-six: twenty-three from mom and twenty-three from dad. Fruit flies have eight. Potatoes have forty-eight. Don't be fooled by the name. Chromosomes are white. They got their name—chromo, the Greek word for color—when, in early genetic research, dyes of different colors were used as means of identification. Each human cell, other than blood and gamete (sex) cells, has four sets of its thirty thousand genes held in the two pairs of double helix. In theory, any one cell of your body could reproduce your entire self. All the information is there. But in any given cell only five to ten thousand selected genes are actively being expressed. This partial expression differentiates one type of cell from another, which in itself is intriguing. How does a cell happen to know where it is so that it produces the proteins that form an ear where ears are to be and an eye where eyes are?

The DNA helix is some 25 billionths of a meter thick, but reaches a tenth of a meter (about 3 inches) in length when unwound. If we were to unwind all the DNA of a cell and line it up, one cell's worth would reach about the height of an adult. Put all the DNA in your body end to end and you get a thread that reaches to the sun and back about one hundred times. This is not some abstract fact. It is the wonder of you.

DNA codes its information using combinations of four base nucleotide molecules. It's the genetic alphabet of life: adenine (A); guanine (G); cytosine (C); thymine (T). In RNA, thymine is replaced by uracil (U), thus obviating the possibility of the cell confusing RNA, the copy of the information, with DNA, the source of the information. Another clever trick by nature that keeps the number of genetic errors low.

DNA consists of a double strand of these nucleotides strung along two spines of repetitively alternating sugar and phosphate groups. Here each base along one spine bonds with its complementary base on the opposite spine: A bonding with T; G bonding with C. (In RNA, A bonds with U.) This bonding twists the two strands one about the other into the famous double helix, the discovery of which in 1953 earned Crick, Watson, and Wilkins the Nobel Prize in 1962.

When information is to be retrieved, the relevant portion of the double helix is unwound and opened. In this configuration, the two spines form parallel lines, with the paired bases coming off at right angles. What resembled a helical spiral staircase has been reshaped to that of a ladder, with the linked bases forming the steps and the two spines the outside handrails. The DNA is now in position to provide the information needed in protein synthesis. (See Figure 17.)

Proteins are strings of amino acids, but genes are strands of nucleotides. Combinations of twenty different amino acids are used in the proteins of life; there are four bases used in the nucleotides. Potential for a language problem exists here. Nature facilitates the translation from bases, similar to the letters of the code, to amino acids, the "words" of the protein (with the proteins being the sentences) by using groups of three bases as a code for each specific amino acid. GCG, for example, codes for the amino acid alanine, CGG for arginine.

The immediate question that arises is: Why did nature bother with the coding to store the secrets of cellular life? Since at any moment a cell is organizing amino acids at the incredible rate of thousands per second, why not store the information in the nucleus directly as a string of amino acids, or have twenty different baselike molecules, each one indicating a specific amino acid, rather than the three bases that nature actually uses? My hunch is that thirty thousand genes are simply not adequate instructions to produce a human from an egg. The complex chain that leads from DNA to RNA to protein allows for the string of nucleotides to contain information other than merely the genes. If so, then the amount of information able to be contained on these chromosomes can be vastly expanded. We are no longer limited to the genes per se. The multiple levels by which each aspect of the physical world can be read, such as the wave/particle and energy/matter dualities, may find parallels in the reading of the genome.

When a specific protein is needed by a cell, a chemical messenger is sent from the outer cell, through a pore in the nuclear membrane, into the nucleus. How the messenger knows to go to the nucleus currently remains a mystery. This messenger finds the needed chromosome (one of the twenty-three pairs), locks onto that chromosome, and moves along, nucleotide by nucleotide,

DNA/RNA: THE MAKING OF A PROTEIN

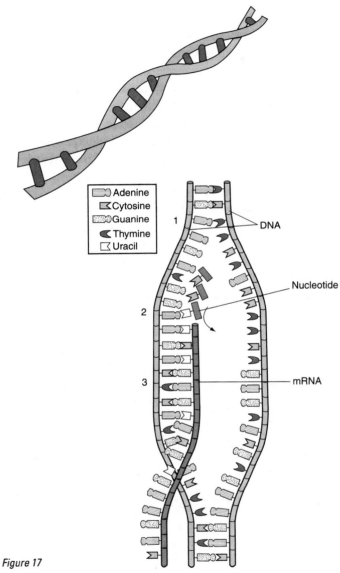

Legend:
- Adenine
- Cytosine
- Guanine
- Thymine
- Uracil

1 — DNA

Nucleotide

2

3 — mRNA

Figure 17

Synthesis of mRNA

Matching and joining of the nucleotides to form the mRNA takes place at the rate
of 50 substitutions per second, a feat of molecular gymnastics.

until it comes to the specific sequence of bases that marks the beginning of the gene that codes for the desired protein.

At this stage, the signaling molecule changes shape, and in doing so allows—or causes—an enzyme called DNA-dependent RNA polymerase (I'll call it RNA-P) to join the action. In fact it really starts the action. This amazing molecule is a phenomenon in itself. It binds to the promoter region of the chromosome situated at the start of the gene and at that point breaks the hydrogen bonds holding the parallel strands of DNA (recall that these bonds are strong enough to pull the DNA into its helix spiral, but not so strong that they obviate the rapid opening of the rungs on the DNA ladder). Once the helix is opened, the nucleotide bases are ready to be read. The RNA-P now moves along the exposed nucleotides, identifies each base, and assembles a chain of their complementary bases. For base A, a base U is selected from the surrounding cytoplasm; for G a C is selected and joined to the previously selected U.

To get a feel for the complexity and speed of transcription, please take a deep breath and read the following sequence of events in one burst. It's what your RNA-P is doing inside every cell of your body, a thousand times over, right now. The RNA-P opens the helix, reads each nucleotide base, selects the correct complementary base from among the four types floating in the intracellular slurry, concurrently selects from the same slurry the molecules that make up the spine of the lengthening strand of mRNA being manufactured, trailing behind the RNA-P, joins the just-selected base to the spine, takes the portion of DNA that has been read and reseals it to the parallel DNA strand from which it was separated, opens the portion of DNA to be read next, reads it, and continues this juggling act till it reaches a coded stop order. Exhausting just to describe it. And RNA-P does this manufacturing at fifty bases a second, a lot faster than you read the words I used to express just one step of the fifty. ATP is there to provide the needed energy at each junction. Keep in mind, this entire sequence is performed by molecules reading molecules, molecules selecting molecules, molecules walking along with other molecules. Don't project too much brain power or body power onto the system. It's not little people in there. It's simple molecules that somehow seem to act like little knowledgeable people, as if they had a wisdom of their own. Which they do.

Because the demand for proteins is so great, as many as thirty RNA-P molecules may be at work simultaneously along a single gene, churning out multiple copies of the single-stranded mRNA. Electron micrographs of an active gene show strings of partially completed mRNA extending at right angles from the DNA spine, with increasing lengths of the mRNA as we move along the spine toward completion of the gene. The impression is something like the branches of a balsam fir seen in silhouette, short near the start of the gene and broadening as we approach the end.

With the mRNA now copied from the DNA, a further complication arises. If DNA is 90 percent "junk," then so is the complementary form of mRNA. This "junk," known as introns, must be removed, snipped out by some very wise snipper and discarded. Then the remaining sections, known as exons, the sections of bases that code for the wanted amino acids, must be spliced together by a very clever splicer. With that completed, the mRNA is ready for its task, making a protein.

Through a process not yet understood, each completed, snipped, and respliced mRNA is transported through a nuclear pore and into the outer cell cytoplasm, where two parts of another of the cell's wunderkind, the ribosome, join together, clasping the mRNA within. The business of a ribosome is to read the information sequestered on the mRNA in groups of three bases, call for another type of messenger, transfer RNA or tRNA, which brings the amino acid that corresponds to the just-read three bases (one of twenty types of tRNA in the surrounding fluid, one type for each of the twenty different amino acids used in proteins) to the ribosome and then to join the amino acids one to another, as the bases are read, thus forming the needed protein. Easier said than done. And yet with each of your seventy-five trillion cells incredibly producing two thousand proteins each second, including this second within your body, nature makes it seem easier done than said.

The detail continues. The cellular mechanism literally swims into action. It makes our heads spin and it should. The wonder of life is in the dovetailing intricacy. If the requested protein is for export, manufacture occurs at an organelle known as the endoplasmic reticulum (ER). Here thousands of mRNA are at work making proteins that upon completion are swallowed by the ER, a portion of which then pinches off, forming a pouch called the Golgi apparatus. Golgi are carried to the cell membrane on the backs of motor proteins (molecules that actually walk along the microtubule tracks that lace the cells and know where to and when to walk). When signaled, the Golgi fuse with the membrane, pop open on the outer side, and spew their contents of newly made proteins into the extracellular fluids and blood. I wonder just how each Golgi knows which direction to send the motor protein that is carrying it. Marvelous.

And there are the huge chaperone molecules which engulf the entire newly made protein, check it for errors by examining the protein's charge characteristics, and then, if an error is found, send it back for retrofit. There are a hundred and more functions that in the interest of brevity I have not described, such as shopkeeping, diverse manufacturing, transporting functions, and others. All are going on simultaneously in each cell, each function equally crafted like a fine Swiss watch. We take it all for granted. We see the system from the outside. But from the inside, like turtles, it's a wonderland all the way down.

Five minutes ago I couldn't even spell "endoplasmic reticulum," and now I know I'm full of it.

MITOSIS AND THE

MAKING OF A CELL

"Acquire wisdom and with all your resources acquire understanding." (Proverbs 4:7)

In the text we skipped one major step in the making of a person. We covered meiosis, the sexual reproduction that starts the process. But immediately thereafter, a fetus needs a plenitude of proteins sequestered in a universe of cells. Lots and lots of cells. In fact, about seventy-five trillion for an adult. And they all start from one single cell that somehow knows to divide, and to differentiate into a variety of cell types, and to divide again.

Aside from the brain, biological cells are probably the most complicated entities on the face of the earth, and possibly, hubris notwithstanding, the most complicated entities in the universe, far more complicated than we would imagine from an external view of the organisms they combine to make. The complexity of life's biology is held within its internal processes, but dressed in the simplicity of our outer bodies. Anyone who thinks nature is simple should take a short user-friendly course in how just one cell "evolves" into two cells, and then marvel at the vast amount of information required to bring about this single step and then wonder at its source. Let's have that course now. The hidden face of God is found in the details.

Packed into a membranous ball a few tens of a millionth of a meter in diameter are close to ten thousand proteins and another thirty thousand molecules, each busily occupied with its own intricate functions, processing the information encoded for life. Information is the key ingredient that shows up in every aspect of life.

To reach the 75,000,000,000,000 (i.e., 75 trillion) cells in an adult body, only forty doublings of the cells would be required if all the cells divided at an equal rate. That is, 75 trillion is about 2^{40}, two to the power of forty, two multiplied by itself forty times. The entire process of cell division, referred to as mitosis, takes only about an hour. Which might imply that if we can move from fertilized egg to full adult in forty generations of cell divisions, the entire process should be finished in a mere forty hours. But it isn't so easy. Between each session of mitosis, a cell goes through an interphase lasting some ten to twenty hours, depending upon cell type, during which (among other activities) the cell is occupied with protein synthesis—that amazingly complex flow in which individual amino acids are brought into the cell, strung onto specifically ordered chains, snipped, respliced, and then shaped into structures of life.

Still, even at twenty hours per division, forty generations would occupy a mere eight hundred hours—about a month. But of course, twenty years is closer to the reality in the making of an adult member of our species. So what's taking so long?

First, most cells don't live for the entire life span of the body. They age, die, and must be replaced. The vast majority of newly manufactured cells are of this replacement variety. Hence the total tally of cells present at any one time pales in comparison to the numbers that have actually been, or are being, made.

Right now, within your body, new cells are being produced at a clip of four to five million this very second, and every other second, too. A typical cell, 20 to 30 microns (millionths of a meter) in diameter, has an approximate weight of a billionth of a gram. At four million new cells per second, that equals four milligrams of cell weight each second, 400 grams a day, or about 140 kilograms (300 pounds) of new cells produced each year! No wonder it is hard to keep our weight in check. But don't blame it on the cells. It is safe to say the you that was you a year ago, the atoms and the molecules of your body, is not the you you are today. Your body sloughs off, discards, close to the entire 140 kilograms of body tissue each year. Epidermal skin (the outer layer) is continuously being replaced. These cells last about a week before being traded in. In your mattress there's an entire ecosystem of micro-mites eating what was you yesterday.

Red blood cells account for 40 to 50 percent of total blood plasma. They are packed with oxygen-carrying, iron-rich hemoglobin—hence the red color of the oxidized iron—and live for about one hundred days. Almost a million die and are replaced each second, the product of our red bone marrow. Cells in the less user-friendly environs of the stomach and intestinal lining, where the acidity (pH = 1) is ten times that of lemon juice and a hundred times that of vinegar, coke, or beer, survive only a day or two. Intestinal juices will put a pucker on your lips if you taste them. (Blood is about neutral as far as acidity is concerned; pH = 7.)

Some cells, such as nerves and skeletal muscle, in theory can survive for the entire life of the individual if not damaged, which sounds good but has concurrent disadvantages. A severed nerve, unable to be replaced, may result in permanent loss of function in the related muscle. While muscle cells do not divide, damage triggers a process that can reconstruct the injured tissue.

In addition to the demise of cells as a limitation on mitotic accumulation, cellular metabolism plays a key role. In the life cycle of essentially every biological cell, there is a phase or stage known as GO. Contrary to what the name might suggest, the GO phase signals the cell's exit from the reproductive flow. While continuing to be metabolically active—that is, producing enzymes, hormones, and other proteins upon demand—a cell in the GO phase becomes nonproliferative, extending the entire mitotic cycle well beyond the ten-hour minimum.

Part of the tragedy that leads to cancer is an inopportune mutation in a chromosome that signals a cell to bypass the GO phase. Since cell division involves replication of the chromosomes, the mutation is passed on to the cell's progeny. Generation by generation, the number of these wayward cells increases, and each one replicates without pause. Without GO to extend the duration of the cycle, mitosis occurs each ten hours. Within four days the errant cell has laid the basis for a thousand cells, and in eight days the tally is a million. If the cancerous cells can spread from their initial site, they soon tap the resources of the entire body in their unchecked lust for growth and reproduction.

Errors in the genetic code can arise from a variety of causes, including a simple mistake in copying the DNA during its replication. Hence cell types that divide frequently, such as skin and stomach, have high incidences of carcinoma. Cancer among long-lived muscle and nerve tissues is far more rare. Some diets correlate positively with cancer. Fat intake, for example, correlates with breast cancer. It tends to occur once for every ten women who consume on average 150 grams of fat per day, but has almost no occurrence in those rare populations consuming less than 40 grams of fat per day. Diets rich in smoked and pickled foods correlate positively with stomach cancers.

As we progress in our journey through the complexity of a cell's reproductive cycle, reflect upon the fact that within each healthy adult body, cell division is happening at a rate of four or five million times per second. Don't lose sight of the fact that you represent an awesome wonder.

The most crucial stage in cell development, the actual replication of the genetic material, occurs long before any externally visible change in the cell wall is noted. This came as a surprise to biochemists, since logic would have it that the entire mitotic process would be one integrated affair. Not so. Though the duration of total cell cycles varies widely among tissue types, the time required for the actual mitosis, the division of the parent cell into two daughter cells, is similar for all—about one hour.

If we take sixteen hours as a typical cycle for the entire metabolic procedure in cell growth and function, DNA replication starts a full ten hours prior to the mitotic division of the cell. The process is quite similar to that which we encountered in RNA production.

The primary factor facilitating the accurate replication of the long molecules of DNA is that each of the four nucleotide bases will bind properly only with its complementary base: A with T, G with C. The chemistry behind this exclusive pairing serves as a spell checker: T and A join together with two hydrogen bonds; G and C bind with three hydrogen bonds. If by error a T binds to a G, the G is left with one hydrogen unpaired, a signal that an error has been made.

Proteins, named DNA polymerase (I'll call it DNA-P), enter the DNA double helix, unravel a portion of the helix, separate the strands, and then move nucleotide by nucleotide along each of the two now-separated chromatin (DNA) fibers. On each fiber, the DNA-P draws from the surrounding slurry of the cell's nucleus the nucleotide base complementary to the nucleotide on the parent chromatin strand and then inserts it onto the newly molded chromosome skeleton, the backbone of the DNA strand, the string of alternating sugars and phosphate molecules that is also being simultaneously formed. Since each of the two newly formed DNA strands is a copy of the older strand of DNA that previously occupied that same position, as the DNA-P moves along the parent DNA strand, the new daughter strand mates and helically winds with the original parent. It sounds almost oedipal.

Unlike RNA transcription, where only one strand of the

DNA is copied, in cell division both strands of the double helix are being copied simultaneously. Therefore, upon completion, the cell has two complete sets of chromosomes, or ninety-two in all.

The cleverness of the system is subtle. Not just how it learned to get a protein to open the helix, or how DNA-P is made and then knows to come on the scene, or how it finds and joins the correct base. Those acts in themselves are near-wizardry and plead for explanation. The cleverness here is that each new strand winds helically with the parent from which it was, as a complement, copied. Brilliant! Now quality control proteins can check the new work directly against the original template, the parent strand. Had the two new strands wound about each other, new to new, and the two parent strands re-formed their original helix, the quality control would have been a far more difficult and far less efficient task. With the new DNA bound to the old, if the quality control protein finds for example a base A in the new strand at a site where in the parent there is a G base, it "knows" an error has been made, and it knows the error lies in the new strand. So it clips out the base A, draws a base C from the slurry, and splices it into the strand. The cleverness of these proteins sounds almost human, but they are only molecules that together can make a human body.

In DNA replication, the DNA-P reads about fifty nucleotides per second, the same rate as in mRNA production. Even at that amazing clip, considering that there are seven billion nucleotides that must be copied, we're talking a very long time to complete the project. Seven billion copies divided by fifty copies per second equals 140 million seconds. With about 30 million seconds per year, we need over four years to copy one cell! Nature gets the job done in about ten hours by having a few thousand DNA-P enzymes working simultaneously, each copying a portion of the parent, and then splicing the parts to make the whole chromosome. In human terms, the feat of completing the copying job in ten hours is as if a team of readers would plow through ten books, each four hundred pages long, every minute, nonstop for the ten-hour period and, while reading, would simultaneously organize and coordinate the information into a single coherent text.

We've made the double load of DNA bursting with potential, all sequestered in a single cell's nucleus. The cell waited patiently ten hours or so as the bevy of DNA-P feverishly turned out the second copy. Notice that though there are thousands of DNA-P enzymes working in unison, we don't get, let's say, forty-eight pairs of chromosomes or forty-two pairs. The tally always comes out just right, forty-six pairs. They seem to know not to re-recopy that which has been already copied.

Mitosis can now begin. And once it starts the process is non-stop activity until the end. In the hour that it takes, it is fantasy brought to life, pure and simple. Disney could not have done it better. Forget religion, forget theology, universal consciousness, all the esoteric stuff. Just immerse your mind in the awesome biological reality of you.

Outside the cell's nucleus (which now houses a double set of the chromosomes), two organelles referred to as centrioles migrate to opposite sides of the cell. The membrane surrounding the nucleus starts to disintegrate and at the same time the centrioles begin organizing microtubules, each some 25 billionths of a meter in diameter, into spindle fibers that extend between what is shaping up to be two opposing poles of the cell. Keep in mind, something has to cue this synchronous dance of the molecules. Somewhere inside the cell there's a molecular mind in tune with a molecular clock.

While this is happening, what appeared at first to be a spaghettilike jumble of the ninety-two chromatin strands condenses into chromosomes that upon close examination are actually twenty-three pairs of pairs, that is forty-six pairs or ninety-two chromosomes in all. Each pair is joined at a single point. Some pairs appear as Xs and some as Vs, depending upon the location of the constricting bond. Each pair consists of two identical copies of a given DNA/chromosome. Now that might represent a problem. Cells don't want double pairs. They want single pairs. And that is what mitosis is all about. (See Figure 18.)

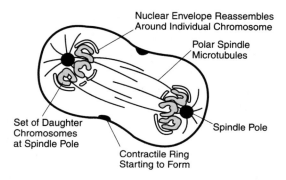

Figure 18

Telophase in Cell Division: The Fifth Stage of Mitosis (Figure after B. Alberts et al., *Essential Cell Biology,* Garland Publishing, New York, 1998)

Some thirty-plus minutes have passed since the start of mitosis.

The spindle fibers, extending from the poles of the cell in a manner not totally understood, now draw the chromosomes to the central region of the cell, setting them in a line perpendicular to the fibers that pass between the poles—something like a map with a central equator made of the chromosomes, laced with fiber lines of longitude running to the north and south poles. Spindle fibers have bound themselves to each chromosome at the point of constriction. And then, wonder of wonders, miracle of miracles, a motor protein associated with each chromosome grasps with its little molecular hands the associated spindle fiber, contracts pulling its particular chromosome pole-ward, releases, reextends its hand, grasps and contracts again in a manner similar to the functioning of muscle proteins, each stage pulling its burden a bit closer to the pole. Tugging, straining, reaching out again, and pulling some more until finally the single link at the point of constriction breaks and the sister chromosome pairs move apart, each toward its respective pole. Five to ten minutes pass as the motor proteins reel in their cargo through the ten or so microns of cytoplasm to reach the pole. As always, energy-rich ATP is there to power each molecular effort.

About forty minutes have passed since the start of mitosis.

With the chromosomes separated into two identical sets of twenty-three pairs each, one set at each of the poles, the final phase begins. New nuclear membranes form, individually packaging the two sets of genetic material safely away from the ensuing activity. The helical chromosomes can now unwind, exposing their genes for reading. At the equatorial line of the cell, a ring of muscle protein forms. This, as with all muscles, consists of fibers of actin and the motor protein, myosin, with its hands that grasp, contract, release, extend, and grasp again. As the myosin pulls the actin ring ever tighter around the cell's middle, the cell membrane begins to contract at what was formerly the equator. The cramping continues and ultimately the cell divides in two, with the original cytoplasm and the organelles therein being shared between the two halves.

One hour has passed since the start of mitosis.

Over the next five or so hours, the identical genes in each of the two new cells will produce more organelles (ribosomes, mitochondria, ER, etc.), making each of the cells full fledged and full sized. The cells now have the choice of continuing toward another round of division or entering the GO phase of nonproliferative metabolic activity.

But there is a bit more to the story. At two stages, one before and one after the actual mitotic phase, the cell checks itself for overall size and DNA functioning. Then, during the one-hour mitotic phase, the spindle fiber network is monitored, determining if the fibers have joined properly to the chromatids. Any malfunctions are repaired. If not repairable, the cell self-programs for cell death and disintegration. Its usable parts are recycled in other cells.

That's how we work, but how did we evolve? Not through random reactions over billions of years in energy-rich ponds, thermal or otherwise. The fossil record has laid waste to that false assumption, so popular prior to the discovery of the earliest fossils. As so many secular scientists active in the search for the origin of life have expounded: The data indicating the rapid emergence of life on our initially lifeless planet teach that life is inevitable in our universe. True enough. Life seems inevitable. The question is, what made it inevitable?

If life is inevitable, I have to ask why? But try as I might, I cannot envision this complexity evolving without powerfully guiding restraints and catalysts propelling it toward completion from its beginning. Call those restraints and catalysts nature if you like. But it is an amazing nature operating in a manner never dreamed of as being naturally possible just a few short decades ago, a time when life seemed to function with an almost childlike simplicity. Then came the discoveries of molecular biology, the cellular mechanics of life, that allowed us to peer beneath the veneer of simplicity and discover the depth of information sequestered therein.

MUSCLES: MOLECULES

IN MOTION

The text covers the complex orchestration of cellular processes that sends a nerve's signal careening down its axon. This wonder of biochemistry is matched by the intricate functioning of the muscles that those signals set into action. Muscles provide life with the option of motion. They account for almost half of an adult's body weight and use close to a quarter of the oxygen that we breathe, even when we are at rest. During exercise, oxygen consumption in some muscles can increase greater than tenfold. Studying muscle dynamics can be as breathtaking as using them.

Humans have three basic categories of muscle: smooth and involuntary, such as those that line the walls of blood vessels and intestines (no conscious thought is associated with their pulsing); cardiac, which are also involuntary for all of the two to three billion beats of a lifetime, but are structured in a manner similar to the third muscle type; and skeletal, which provide the voluntary motion we usually associate with muscle building.

It seems paradoxical that although we flex our limbs in and out, up and down, muscles produce force only by contracting, by pulling in. They pull, but never push. By the brilliant design of joints and levers in our limbs we can do "push"-ups. And perhaps equally surprising, when a muscle fiber gets the message from a motor neuron to contract, the fiber can only contract fully—never partway. Yet the wisdom-packed design of muscle fibers clustered into woven ropelike groupings enables us to crook a finger partway, to open our mouth a bit to whistle and a lot to bite. Does design of this level imply a designer? Some very smart persons say yes; some equally smart persons say no. After decades of research in the physics and biology of the world, I opt for yes as the answer to that question.

All three classes of muscle, indeed all muscles in all eukaryotic life, function in similar fashion. The main players are cables of actin and filaments of myosin arranged in overlapping parallel fashion that pull against one another—something akin to the myosin shimmying up a rope of actin. Myosin is equipped with hands with a powerful grip that swing forward and grab the actin. By bending its "wrist," the myosin pulls itself forward along the actin. That's it in brief. The details hold the wonder, and to me at least, another indication of "the Force" active in the system.

The molecular chain known as actin forms in pairs that twist into a double helix (as does DNA). Each twist of the helix is a mere 70 billionths of a meter in length and approximately 10 in diameter. Binding sites for the myosin are closely spaced along the actin molecule, but due to the helical twist of the two woven molecules, some of the potential sites are masked. The result is an exposed actin site that myosin can grab only ("only") every 3 billionths of a meter, more than enough for the needs of a body.

Myosin wants to grab. So the impulse must be regulated, controlled by our free will and choices that in turn control the neurons that signal for muscle contraction. The regulation is accomplished by a pair of proteins, one of which blocks myosin access to the actin-binding sites. The second binds to the first and has a site for calcium ions. The action potential of a nerve signaling a muscle to move into action reaches its axon terminal and causes the Golgi apparatus to release its neurotransmitters into the nerve/dendrite synapse. These, in milliseconds, diffuse across the synapse and bind to receptor molecules on the mus-

cle fiber's outer membrane. The signaling nerve's neurotransmitters, upon binding to a muscle fiber, stimulate calcium ion
channels in the muscle cell membrane to open. Calcium ions
flood into the cell and bind to the receptor sites on the secondary blocking protein, causing it to flex. In the flex, it pulls back
the primary blocking protein and, in doing so, exposes the actin
sites. Myosin makes its grab.

Is this a Rube Goldberg machine? Couldn't nature just as
simply have had a nerve send a signal to the myosin telling it
when not to grab and when to grab the actin and contract the
muscle fiber? Perhaps in this complexity we are seeing the
structures that led to the system we see today, something like
the scaffolding required to support an arch while the arch is being constructed. Could the complexity be evidence of evolution,
rather than design?

One could imagine the "evolution" of an arch. First there'd
be the need to span a space between two pillars or two ridges
over a narrow channel. A straight, horizontal beam between the
two, each end resting on a side, would do as a starter. Then, to
gain more height, two beams set in a V-shape, but pointing upward, might accomplish the goal. In time one could envision
adding further pieces to smooth over the sharp "V" vertex and
add strength. This could eventually allow the wood pieces to be
replaced by an arch. In this scenario, we avoided the severe snag
of first having to "evolve" the complex biomolecular mechanisms of photosynthesis that produce the cellulose and the simpler carbohydrates and then from them "evolve" the tree from
which the necessary beam was taken. We took the tree as a
given. In its development or evolution, the tree trunk grew in
order to span space, to allow its leaves to rise above the forest
canopy. We merely adapted that vertical spanning feature for
our horizontal needs.

With an arch, once completed, the scaffolding is removed;
no trace of it remains. Nature tends to be conservative, keeping

the "scaffolding" in one form or another, such as the yolk sac or the one-chambered heart or the primitive tail we all had while in the womb. So if muscles did evolve, their mechanism should show signs of the scaffolding—the equivalent of old pieces of wood. If the scaffolding is there, it certainly is not obvious.

The myosin has a shape reminiscent of the letter "r," but with the rodlike shaft much elongated. The rods bind together to form myosin filaments, each filament having several hundred protruding hands. The hand is composed of two globular molecules that between them grasp the actin site. In the power stroke, the hand binds to an adjacent actin site, and then bends down, pulling the myosin filament along the actin cable, causing the actin and myosin to slide by each other. The visible result of this phenomenon is the bulge of the bi- and triceps when we flex our muscles. As the fibers slide along and the muscle contracts, by nature, the muscle must bulge out to contain the contracted fibers in the shortened muscle length. Each stroke moves the myosin approximately ten billionths of a meter. Each hand can cycle some five times per second.

I say the hand of the myosin bends down, but that hand is not like ours, having a supple wrist and distinct skeletal parts. Here, with myosin, the hinge is one biomolecular unit, a molecule that is flexible. It actually bends upon command, the result of just-right electrical charges pulling among the individual atoms. When working in unison with thousands of other myosin heads, the pull can have a pair of biceps raise a three-hundred-pound weight, and yet not have the force at any one of the heads exceed the strength of the actin molecular bonds, which would tear the cable.

Each muscle cell consists of thousands of fibers, each of which contracts upon stimulation by a nerve. Transverse tubules carry the signal to the individual fibers. In synchronous action, some myosin hands grasp the actin and pull, while others are releasing their grip and reaching further along the actin cable. On a nanoscale, it's one hand over the other, pulling along a rope. The combined effect is a continuous and smooth flow of muscular power. Once the nerve signal ceases, the calcium is pumped out; the muscle relaxes and extends, ready for its next task.

Calcium ions flood in; ATP, manufactured from glucose, is on site to power the myosin; neurotransmitters bridge the synaptic gap. All these materials and a myriad of others I have not mentioned must be on site and on time. And in a healthy body they are. Not by chance, surely.

To move the needed materials from the place of their manufacture to the location at which they are to be utilized, our bodies are provided with a type of muscle quite different from those just described, a muscle not much larger than a molecule. It is the motor protein.

Most materials produced by a cell are made in the cell body. Within these compressed dimensions transport by diffusion is adequately fast. In 50 milliseconds, diffusion can spread a product about 10 microns, a typical path within the cell. But in a nerve cell, with its elongated axon, the point of need may be a meter distant from the point of production. If a particular protein required at the terminal were to travel from cell body to terminal by diffusion, a year would not suffice to have it complete the journey. Clearly some means of active transport is essential. Nature has provided both the vehicle and the pathways, the vehicle in the form of motor proteins and the pathways as microtubules that run the length of the cells.

Motor proteins, approximately two billionths of a meter across (almost ten times smaller than the actin/myosin muscle units), move along the microtubule tracks in pairs. The lead motor protein locks onto the microtubule thrusting forward, and in doing so pulls free the trailing motor protein, which then seems to leapfrog to the lead position. This action is repeated as the pair travels along the microtubule. On a molecular scale, these two proteins appear to be walking on a rail. Sites on their backs provide a bond for the produce that is to be transported. It is a micro-world of traffic. As with muscle motion, ATP is the power source that allows the motor proteins to leap one past the other.

At a transport rate of 20 to 40 centimeters per day, two days and more is required to move Golgi apparatus housing neurotransmitters from cell body to axon terminal. It seems the cell must plan for events several days in advance. The neurotransmitter packaged within the Golgi apparatus somehow finds and mounts a motor protein headed not by chance down the very track that will bring it to the axon terminal two days later.

Our cells are a nonstop marvel. Transport in all directions satisfies the needs for the two thousand proteins manufactured every second of every day, seven days a week. No nighttime snooze and no Sabbath rest here.

Muscle distribution within our bodies is filled with cleverness. Hold your hand up and bend your fingers. Notice that the muscles that allow you to cup your hand by bending your fingers down are not located in your fingers. Make a fist and feel the inner, smooth side of your arm just below the elbow. Feel those muscles flex. They are connected via tendons to your fingers and give the pull that shapes your fist. By having the muscles located on the arm rather than at the fingers, the fingers remain slim enough to do fine work such as holding a stick or typing a page. But when you pull your fingers down, there is another joint in the line of action, the wrist. Why isn't that pulled down along with the fingers? Now feel the outer, hairy side of your arm just below the elbow. Feel the other muscles at work there. They get the command to apply just the correct force to hold the wrist steady when your brain says bend fingers only, and they allow the wrist to bend when the cranial message is: wrist also in action. But we never think of it because it's all controlled at the less-than-conscious level.

There's a brain of which we are conscious and one we are not. Just as there is a world we perceive and one we do not. Both are real. And with careful thinking we can realize the presence of both.

ACKNOWLEDGMENTS

Prior to the start of my work on *The Hidden Face of God*, I had the vague feeling that some commonality might pervade all existence. With the research for, and the writing of, this book the extent and manifestation of that universal oneness has become for me a part of my daily experience. Many persons contributed to this realization, some through sharing a brief moment of inspiration, some in many hours of conversation, some by their writings. I am greatly indebted to all. Among these are ordained theologians, persons with advanced academic degrees, winners of prizes in science and philosophy. I list here only the names, without titles.

First my wife, Barbara Sofer Schroeder, and our children, Hanna, Yael, Hadas, Joshua, and Avraham, provided plenty of food for thought in our many dinner table discussions.

As with *The Science of God*, Debra Harris and Beth Elon brilliantly directed the manuscript to Bruce Nichols, editor at The Free Press. Bruce's discerning editing helped rid the wheat of the considerable chaff I had initially included. Edith Lewis helped put the finishing touches on the style. As in the past, I am indebted to Helen Rees and then Marc Jaffe and Michelle Rapkin who, prior to the publishing of *Genesis and the Big Bang*, realized the potential for interest in a rigorous approach to the science/Bible debate.

Sincere thanks to Noah Weinberg, Yaakov Weinberg of blessed memory, Dennis Turner, Avraham Rosenthal, Shmuel Silinsky, Motty Berger, Sam Veffer, Nadine Shenkar, Ari Kahn, Zola Levitt, Sandra Levitt, Paul Joshua, Benji and Leah

Schreiber, Cedric Levy, Peggy Ketz, Helen Stone L'or, Mordechai Geduld, Aryeh Gallin, Susan Roth, Ilana Attia, Michael Behe, Lee Spetner, Michael Denton, Moshe Schatz, Michael and Karen Rosenberg, Sharon Goldstein, Nancy Sylvor, Naomi Geffen, Barry Bank, Marty Poenie, Jonathan and Elaine Sacks, David Lapin, Beril Wein, D. Homer Buck, Michael Corey, Phil Rosenbaum, Yigal Bloch, Barbara S. Goldstein.

The books and articles that provided information and insights were *Human Physiology* by D. Moffett, S. Moffett, and C. Schauf (Mosby Publishing, St. Louis, Mo., 1992); *Essential Cell Biology* by B. Alberts et al. (Garland Publishing, New York, 1998); *Essentials of Genetics* by W. Klug and M. Cummings (Prentice Hall, Upper Saddle River, N.J., 1999); *Mapping the Mind* by R. Carter (Weidenfeld & Nicolson, London, 1998); *Tour of a Living Cell* by C. de Duve (Scientific American Books, New York, 1984); "The Origin of the Universe," eds. H. Branover and I. Attia, *B'or Ha'Torah*, Number 11, 1999; *The Fifth Miracle* by Paul Davies (Simon & Schuster, New York, 1999); "Facing Up to the Problem of Consciousness" by D. J. Chalmers, *Journal of Consciousness Studies*, 1995; *The Natural History of Creation* by M. Corey (University Press of America, New York, 1995); *Incredible Voyage* (Book Division of National Geographic, Washington, D.C., 1998); *God and the Big Bang* by D. Matt (Jewish Lights Publishing Co., Woodstock, Vt., 1996); *Biochemistry* by L. Stryer (W. H. Freeman, New York, 1995).

INDEX

Abel, 10, 176
Abraham, 14
Actin, 212, 213, 215
Action potential, 95, 98, 99, 101, 102, 109, 162
Adenine, 195, 197, 205, 206
Adrenaline, 64
AI (artificial intelligence), 147–148
Ain od, 13–14
Akiva, Rabbi, 14
Alanine, 196
Amacrine cell, 85
Amino acids, 58, 62, 150, 189, 196, 199–200, 202
Amygdala, 116, 117, 120, 122, 126, 131, 133, 137
Arginine, 196
Atoms, 3–4, 6, 7, 30, 50, 51, 56–58, 62, 170
ATP (adenosine triphosphate), 66–68, 75, 76, 95, 192–194, 198, 209, 215
Autism, 131
Axons, 90, 95–98, 101–103, 108, 109, 127, 130, 133, 134, 137, 138

Balance, sense of, 122–123, 125
Barghoorn, Elso, 51
Basal nuclei, 132–133
Beauty, 16, 17, 23
Be'raesheet, 49, 178, 187
Berra, Yogi, 166
Beryllium, 56
Bible, 12–14, 48, 90, 107
  creation in, 10–11, 46, 49, 53, 58–59, 87, 138, 145, 174
  monotheism and, 21
  Sabbath and, 181
Big bang theory, 41–45, 55, 149, 186

Biology, 22, 48, 50, 60–63, 87
Birth, 86–87
Black holes, 41
Blastocyst, 80, 81
Blood pressure, 131
Brain, 5, 105–128
  amygdala, 116, 117, 120, 122, 126, 131, 133, 137
  brain stem, 121, 122, 124, 131, 132
  cerebellum, 90, 121–122, 124, 126, 132, 134, 139, 141, 144, 166
  cerebral cortex, 5, 108, 110–112, 117, 122, 131–133, 166
  cerebrum, 122, 132–134
  development of, 110–113
  frontal lobes, 113, 122, 131, 134–138, 142, 143, 156, 165
  gender and, 144–145
  hemispheres of, 116, 124–125, 134–135, 141–144
  hippocampus, 122, 126, 140, 141
  illusions, 160–172
  left hemisphere. *See* Left brain hemisphere
  limbic system, 112, 117–118, 123, 126, 131, 137
  mapping of, 6
  memory and, 117–118, 138–141
  /mind interface, 2, 3, 22, 146–159
  parietal lobes, 122, 134, 135
  pons, 131
  principal organs of, 122
  right hemisphere. *See* Right brain hemisphere
  structure and function of nerves, 94–103, 108, 203
  temporal lobes, 131, 134, 140, 143
  thalamus, 115–117, 118, 120, 122, 124–126, 131, 133, 162

219

Brain (cont.)
   vision and, 6, 16–17, 40, 82–86, 92,
      113–121, 124, 135, 143, 151,
      161–170
Brain stem, 121, 122, 124, 131, 132
Breast cancer, 204
Breathing, 131
Broca, Paul Pierre, 142
Broca's area, 142
Broglie, Louis de, 7, 38, 185, 228

Cain, 10, 176
Cambrian explosion, 121
Cancer, 204
Carbohydrate, 66–67
Carbon, 6, 17, 35, 44, 51, 52
Carbon dioxide, 67
Carter, Rita, 94n
Cells, 49–51, 55, 59–69. See also DNA
      (deoxyribonucleic acid)
   components of, 61
   daughter, 70–72, 74, 77–79
   GO phase, 203–204, 209
   meiosis, 71–74, 77–78
   mitosis, 70, 71, 78, 79, 201–210
Central nervous system (CNS), 110
Central sulcus, 134
Centrioles, 207
Cerebellum, 90, 121–122, 124, 126,
      132, 134, 139, 141, 144,
      166
Cerebral cortex, 5, 108, 110–112, 117,
      122, 131–133, 166
Cerebrum, 122, 132–134
Chalmers, David, 5
Chaos theory, 156
Chemistry, 35–36
Chlorine, 97
Chordata, 124
Christianity, 10, 21
Chromosomes, 71–74, 77, 194, 196,
      206–209
Cilia, 75
Cognitive dissonance, 100, 186
Computers, 147–148, 151
Conception, 74–75
Cones of eye, 16–17, 83
Consilience (Wilson), 36
Coriolis, Gaspard de, 55
Corpus callosum, 122, 134, 144
Cosmology, 2, 149

Creation of universe, 2–5
   in Bible, 10–11, 46, 49, 53, 58–59,
      87, 138, 145, 174
Crick, Francis, 47, 53, 68, 195
Crossing over (during meiosis), 72
Cybernetics, 47
Cytoplasm, 61, 74, 77, 79, 150, 199
Cytosine, 195, 197
Cytoskeleton, 65–66

Darwin, Charles, 91, 113
Darwinian evolution, 90–91, 107,
      120, 153
Daughter cells, 70–72, 74, 77–79, 205,
      208
Dawkins, Richard, 103, 120
Dead Sea, 54, 63
Death, 171–172
Deep Blue (supercomputer), 147–148,
      151
Dendrites, 95–100, 108, 109, 111,
      127, 138
Design, imperfect, 3, 9–14, 84
Determinism, collapse of, 19–20, 185
Deuterium, 42, 43
Deuteronomy, Book of, 12–14, 21,
      187
Differentiation, 41, 78, 81
Digestion, 66–67
DNA (deoxyribonucleic acid), 35, 61,
      67, 70, 71, 78, 81, 95, 98, 150,
      178, 189–200, 205
   -dependent RNA polymerase,
      198–199
   double helix structure of, 47, 195,
      205–206
Double convex lens (of the eye), 114,
      115
Double slit experiment, 8, 155
Dreaming, 169–170
Dreyfus, Hubert, 147–148
Duve, Christian de, 51–53, 99, 180
Dyson, Freeman, 7, 31, 179

Eardrum, 125
Einstein, Albert, 5, 25–27, 38, 47, 55,
      87, 154, 159, 177–178, 181,
      185–187
Electromagnetic force, 29–33
Electron capture, 32

INDEX

Electrons, 4, 5, 26–27, 31, 32, 34–35, 37, 56, 170
Electrostatic force, 29, 30, 33
Elegant Universe, The (Greene), 46
Embryo, 79–82, 110–112
Emotional memory, 117, 126, 133
Endoplasmic reticulum, 61, 200
Energy, basis of, 8
Energy/matter relationship, 25–28
Enlightenment, 36
Enzymes, 79–81, 100, 193
Escher, M. C., 166
Eternal universe, concept of, 152
Evans, Robley D., 185
Evolution, Darwinian, 90–91, 107, 120, 153
Exocytosis, 99
Exodus, Book of, 174, 177, 187
Exons, 199

Fallopian tube, 75–77, 79
Faraday, Michael, 29
Fertilization, 73–80
Fetus, 79, 82, 112
Fight or flight response, 64, 112, 117
Figure/ground reversal, 166
First Three Minutes, The (Weinberg), 193
Flood, the, 10
Forgotten Language, The (Fromm), 181
Fossil record, 51, 90, 91, 107, 120–121, 124, 152, 191, 193, 210
Fourier, Jean Baptiste, 54
Fourier analysis, 157–158
Free will (choice), 40–41, 138–139, 176
French Revolution, 36
Fromm, Erich, 181
Frontal lobes, 113, 122, 131, 134–138, 142, 143, 156, 165
Fructose, 76

Gage, Phineas, 136–139, 183
Ganglion cells, 85, 86, 115
Gender, brain and, 144–145
Genes, 71–73, 81, 136, 138. *See also* DNA (deoxyribonucleic acid); RNA (ribonucleic acid)

Genesis, Book of, 10–11, 13, 28, 33, 49, 58–59, 87, 138, 145, 171, 176–178
Genetic code, 63
Gilder, George, 106
Glucose, 66–69, 76, 95, 215
God
    biblical description of, 46
    changing relationship with, 14
    definition of, 2
    monotheism, 21
Golgi, Camillo, 119
Golgi apparatus, 61, 98, 99, 109, 119, 150, 200, 212, 216
Gould, Stephen Jay, 103, 120
Gravitons, 4
Gravity, 4–7, 28–29, 32–33, 41, 43, 44, 56, 158
Gray matter (of the brain), 111, 132
Greek philosophy, 40
Gregg, Thomas, 120
Guanine, 195, 197, 205
Guide of the Perplexed, The (Maimonides), 23–24, 177
Gulf Stream, 54–55
Guth, Alan, 41
Gyri, 132

Hearing, 6, 40, 117, 125–126, 135, 136
Heart, fetal, 110, 193
Hebrew Bible, 135
Heisenberg, Werner, 18, 47, 185
Helium, 42, 43, 52, 56
Hippocampus, 122, 126, 140, 141
Hormones, 94, 112, 131, 144
How the Mind Works (Pinker), 94
Hubble Space Telescope, 180
Human chorionic gonadotropin (hCG), 79
Humor, 143
Hydrogen, 6, 17, 42–44, 52, 56, 57
Hypothalamus, 122, 126, 131

IBM Corporation, 147–148
Inflation, 41–42
Inner ear, 122–124
Insulin, 67
Introns, 199
Ions, 97, 101, 102, 109

Isaiah, Book of, 176, 184
Islam, 10

Jerusalem, 15–17
Joy, as sign of reverence, 22
Judaism, 10, 21

Kabala, 11, 58, 59, 87
Kasparov, Garry, 147–148, 151
Kinetochore (centromere), 73
King James Bible, 49
Knowledge, limits to, 18–19
Krebs cycle, 68

Labor (birthing), 86
LaHav, Yael, 144
Language, 135–136, 142
Laplace, Pierre-Simon de, 20
Lateral inhibition, 161–163
Laughlin, R. B., 179
Lavoisier, Antoine-Laurent, 35–36
Laws of nature, 15, 16, 42, 53, 55, 56,
     63, 91, 150, 191
Left brain hemisphere, 116, 124,
     134–135, 141, 142, 144
Le-olam, 13
Leviticus, Book of, 22, 183
Life, origin of, 49–68
Life Itself (Crick), 68
Life-spans, 10
Limbic system, 112, 117–118, 123,
     126, 131, 137
Lincoln, Abraham, 103
Lithium, 42, 56
Locality, 8

Mach, Ernst, 163
Maddox, Sir John, 45, 158
Magnets, 29–30
Maimonides, Moses, 23–24, 177, 187
Mapping the Mind (Carter), 94n
Mark, Gospel of, 21, 22
Mass, 6–7, 28, 37
Massachusetts Institute of Technol-
     ogy (MIT), 4, 9
Materialism, 106, 107, 120
Matter/energy relationship, 25–28

Matthew, Gospel of, 22
Maxwell, James Clerk, 29–30, 33
McLuhan, Marshall, 9
Medium Is the Massage, The
     (McLuhan), 9
Medulla, 131
Meiosis, 71–74, 77–78
Melatonin, 124
Membrane of cell, 63–64
Memory, 117–118, 138–141
Messenger RNA (mRNA), 98–100,
     150, 194, 197–199, 206
Mettler, Lawrence, 120
Microtubules, 61, 73, 150, 207, 215
Miller, Stanley, 58
Mitochondria, 75–76, 95
Mitosis, 70, 71, 78, 79, 201–210
Molecular biology, 2, 47, 99, 101, 104,
     120, 123, 127, 152, 210
Molecules, 30, 35, 54, 57, 58, 62–69,
     71–73, 95, 202, 203. See also
     DNA (deoxyribonucleic acid);
     RNA (ribonucleic acid)
Monotheism, 21
Morula, 80
Moses, 14, 177, 179
Motor proteins, 215–216
Multiple sclerosis, 103
Multiple tasking, 109–110
Muscle cells, 64, 203, 211–216
Myelin sheathing, 102–103, 113, 132,
     137
Myosin, 209, 212–215

Nahmanides, 177
Natural selection, 91, 94
Nature (journal), 45
Neocortex, 111
Nerves, structure and function of,
     94–103, 108, 203
Neshama, 138
Neurotransmitters, 94, 96, 98–100,
     108, 109, 150, 212, 213, 215,
     216
Neutrons, 32, 42, 56, 170
Newton, Isaac, 5, 16, 28, 29, 32, 35
Nitrogen, 6, 44, 52
Nucleosynthesis, 43–44, 56
Nucleus of atom, 4, 32, 34–35
Nucleus of cell, 192, 194

Occam's razor, 46
Oneness, 20–21, 34, 41, 87
Onkelos, 33, 135–136
Optic nerve, 85–86, 92, 113, 115
Organelles, 74, 79, 209
Origin of life, 49–68
Origin of Species (Darwin), 91
Ovulation, 74
Oxygen, 6, 17, 44, 52, 57

Paleontology, 48, 101
Parallel processing, 109–110, 144
Parietal lobes, 122, 134, 135
Parkinson's disease, 133
Particles, interrelatedness of, 8
Particle zoo, 38, 39
Pasteur, Louis, 50, 161
Pauli, Wolfgang, 34
Pauli exclusion principle, 34, 150
Penrose, Roger, 166
Perspective, 167–168
Phospholipids, 63–65
Photons, 4, 16–17, 26, 30, 31, 33, 154, 155
Photoreceptors, 83–85, 92, 114
Pineal gland, 122, 124
Pinker, Steven, 94, 103
Pituitary gland, 122
Placenta, 80, 81, 86
Planck, Max, 26, 27, 47, 158, 185
Planck's constant, 38
Planets, 43, 44
Plato, 184
Pons, 131
Population Genetics and Evolution (Mettler and Gregg), 120
Potassium ions, 97, 101, 109
Precentral gyrus, 134, 135
Prefrontal cortex, 137
Princeton University, 4
Principia Mathematica (Newton), 28
Probabilistic nature of nature, 156
Proteins, 58, 62, 65, 66, 109, 189–202, 212, 215
Protons, 6, 32, 42, 56, 158, 170
Proverbs, Book of, 88, 201
Psalm 22, 10
Psalm 23, 10
Psalm 33, 49, 88
Psalm 42, 180

Psalm 100, 22
Psalm 104, 49
Puberty, 74
Purim, 174–175
Pyruvate, 67, 68

Quanta, concept of, 27, 158
Quantum mechanics, 18, 20, 27, 38, 40, 153, 156
Quantum physics, 4, 7–9, 47, 178, 185, 186
Quantum wave functions, 27, 28
Quarks, 4, 5

Radiation, 42, 43, 158
Radon, 32
Ramón y Cajal, Santiago, 119
Red blood cells, 203
Reductionist study of universe, 55–59
Reincarnation, 23
Relativity, laws of, 5, 25–27, 47, 159, 181, 185
Retina, 82–86, 92, 113, 115, 116, 162, 168–169
Right brain hemisphere, 116, 124–125, 134–135, 141–144
RNA (ribonucleic acid), 67, 189, 194–196, 205
    messenger RNA (mRNA), 98–100, 150, 194, 197–199, 206
    transfer RNA (tRNA), 99, 100, 150, 199

Sabbath, 22, 179, 181–184
St. Jerome, 49
Salam, Abdus, 33
Schaffer, Henry, 120
Science of God, The (Schroeder), 50n, 101, 176
Scientific American (journal), 50, 158
Senses, 40, 90
    hearing, 6, 40, 117, 125–126, 135, 136
    sight. *See* Vision
    smell, 6, 40, 116, 117, 133
    touch, 90, 134
Septuagint, 49
Sexual reproduction, 70–80
Shenkar, Nadine, 22

Shir, Hanna, 138
Sight. *See* Vision
Smell, sense of, 6, 40, 116, 117, 133
Sodium ions, 97, 101, 102, 109
Song of Songs, 14
Sound. *See* Hearing
Sound waves, 125
Speech, 90, 135, 136
Sperm cells, 74–77
Spinal cord, 121, 131
Spindle fibers, 71, 73, 77–78, 208, 210
Spinoza, Baruch, 36
Starbolinski, Alex, 41
Stars, 43–44, 56
Strong nuclear force, 31–32
Subatomic particles, 39, 56, 170
Sulci, 132
Sun, 43, 44
Supernovae, 56
Synapse, 98, 100, 146–147
Synaptic gap, 98, 99
Synaptic terminals, 96, 98

Tabernacle, 179
Talmud, 136
Temporal cortex, 126
Temporal lobes, 131, 134, 140, 143
Testosterone, 112, 144
Thalamus, 115–117, 118, 120, 122,
    124–126, 131, 133, 162
Thermodynamics, second law of, 53,
    58, 59, 65
Third eye, 127
Thymine, 195, 197, 205
Time, 7, 45, 158
Touch, 90, 134
Tour of a Living Cell (Duve), 52
Transfer RNA (tRNA), 99, 100, 150,
    199

Truman, Harry S., 148
Turner, Dennis, 18, 193
Tzimtzum, concept of, 12, 176

Uncertainty, principle of, 18, 20, 47,
    185
Uracil, 195, 197
Uranium, 44

Van Gogh, Vincent, 143
Vertebrates, 101–102
Vision, 6, 16–17, 40, 82–86, 92,
    113–121, 124, 135, 143, 151,
    161–170
Visual cortex, 116, 118–120, 122, 135,
    143, 162, 165
Vulgate, 49

Wald, George, 50, 51
Water, 57, 63, 65, 67
Watson, James B., 47, 195
Wave/particle duality, 38
Weak nuclear force, 31–33
Weinberg, Steven, 33, 193
Wernicke, Carl, 142
Wernicke's area, 142
What Computers Still Can't Do (Drey-
    fus), 147
Wheeler, J. A., 8, 40, 53, 153, 154,
    179
Wilkins, Maurice, 47, 195
Wilson, E. O., 36
Wistar Institute of Anatomy and
    Physiology, 100, 101

Zechariah, Book of, 13